Multidin

The physics of Multidimensional Time
and Human Consciousness

by
Robert Kersten

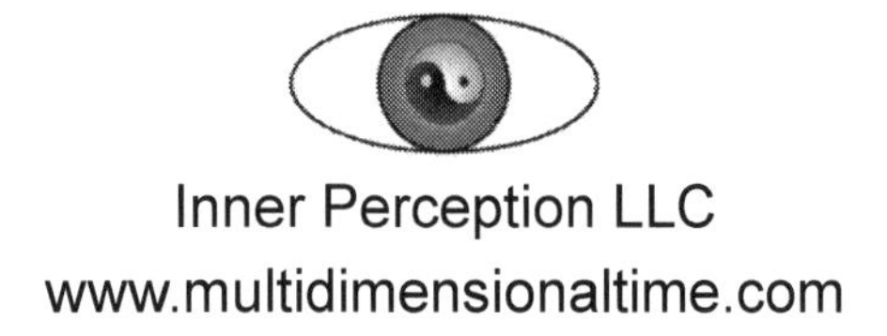

Inner Perception LLC
www.multidimensionaltime.com

Acknowledgements

To my wife Anita, for her support in my continuing journey to understand space and time and her patience while listening to the ideas that I have wrestled with over all these years. To Dewey Larson and Tom Bearden, for making available their ideas on space and time. To Melissa Gray for her copy editing skills. And, finally to Time, my greatest teacher.

To see a world in a grain of sand,
And heaven in a wild flower:
Hold infinity in the palm of your hand,
And eternity in an hour.

William Blake

Relativity

There was a young lady named Bright,
Whose speed was far faster than light;
She set out one day
In a relative way,
And returned on the previous night.

A.H. Reginald Buller

Table of Contents

Introduction

I would rather have questions that can't be answered than answers that can't be questioned.

Richard Feynman

Time is a concept that we think we understand, but if you spend some 'time' studying it—I mean really trying to delve into it—it can make your head spin. It is a subject that seems to defy our ability to explain it in a clear and concise manner. It impacts all of life and science, so the ramifications of a new understanding of time is important.

To start this journey into time, I have chosen to deal with an expanded concept of time in this book, that being three-dimensional time, a postulate that Dewey Larson proposed to explain how matter is formed in a novel way. I found his approach of the speed of light as the link between space and time very insightful. In this book I will explain some of the postulates in Dewey Larson's Reciprocal System Theory, in what I believe is a more intuitive way. I then add to these ideas with concepts from other authors and other areas of science, as well as my own ideas about space and time.

Another physicist and author, Tom Bearden, expressed his ideas of space and time in his book *Energy from the Vacuum.* In this book, he introduces the reader to advanced ideas in electrodynamics, quantum and particle physics. He introduces the idea of a dipole breaking the symmetry of the vacuum and energy pouring back and forth between space and time. His discussions of time, mass, and energy are extremely interesting, and his ideas influenced the way I think about not only time, but electrodynamics and spacetime as well.

While I reference the work of these two gentlemen, I am not saying I agree with all they have articulated, but, in my opinion, their ideas are well worth considering. After a number of reads through parts of these materials over the years, I have integrated ideas of these two physicists along with my own thoughts into new insights into time and spacetime.

Since I am a physicist, I also spent time investigating the implications of these extra dimensions of time to physics, in particular, electrodynamics. Many unsuccessful attempts were made before, a few years ago, I thought I had a reasonable mathematical approach. I wrote a few white papers on my ideas and have included these in the Appendix for those interested in a more technical approach to the potential implications of eight dimensions of spacetime in physics. These papers are also available on a website I setup, www.multidimensionaltime.com. In these papers, Maxwell's equations are evaluated in three-dimensional space and one-dimensional time, as well as three-dimensional time and one-dimensional space. The results bring a symmetry to Maxwell's equations, and the implications are that magnetic monopoles could be found in the three-dimensional time and one-dimensional space, just like electric monopoles are found in three-dimensional space and one-dimensional time.

Like any new skill, getting your mind flexible enough to visualize extra dimensions takes some effort. But with work, the reader can understand and expand on these concepts. The material here is presented to inspire new ideas.

In my own journey, I have worked on these ideas for a period of time, then put them away for all sorts of reasons. But I kept coming back, and *in time*, I found a good correlation between these ideas of multidimensional time and human consciousness. I have always wondered about where the 'I' that felt so strongly about events in my life and in the world around me existed. These extra dimensions of spacetime allow human consciousness to be linked to our familiar spacetime in ways that align with the latest ideas in human consciousness, but now consciousness has a technical basis from which we can understand ourselves in new and novel ways. From these new insights, our relationship to each other, and to the universe, will be supported by these new ideas about space and time.

I hope you enjoy this journey into an expanded multidimensional spacetime, and that it brings a new insight to your journey through your life lived *in multidimensional space* and *in multidimensional time.*

Kind Regards,

Robert Kersten

December 20, 2015.

1. Time

What then is time? If no one asks me, I know what it is. If I wish to explain it to him who asks, I do not know.

Saint Augustine

Saint Augustine's quote at the beginning of this book is a great way to explain our complex relationship with time. We use time to measure the passage of our days, weeks, and years. We schedule school, work, play, and many other activities around this construction of time. When we do this, we feel we understand time. What we often do not realize is that we are using the ***change*** in time to organize our lives. This change is very dependable, since no matter what is happening in our lives, or in our world, this change is consistent. Fortunately.

Our personal experience of time is very different. We find that peak emotional experiences can rapidly change how we experience time. In times of heightened danger, it appears to us that time slows down. In times of peace and joy, time slips by effortlessly. But, commonly, we feel we are in a constant race against time, where time regularly appears to be the winner.

In meditation, people can have an experience where time appears to diminish. The constant noise of everyday life slips away, and a stillness is experienced within. But this effect of time fading away into the background of our consciousness is also experienced by people who are doing activities that they totally lose themselves in. When participating in these activities, whether it be sports, music, writing, or a myriad of other activities, people speak of 'being in a zone.' In this experience, we experience our activities with a grace that produces a peak in performance or creativity.

In spirituality, it is common to talk of a *now* time, a time freed from the burdens and memories of the past and the expectations and anxiety of the future.

This *now* time is the only real moment in time. We can point to our experience in the *now* both inside and outside of ourselves. It is that ever-fleeting present moment that is constantly changing to the next moment, impossible to articulate as it is happening. We can only experience it. The moment consciousness enters in, it was already gone and in the past.

To access any time outside of the *now* moment, we have to go inside of ourselves. To access the past, we reach back into our memories, or we point to previously recorded images or words. But these are only representations of our experience in time, since this time does not exist outside of ourselves.

To talk of the future, again, we have to go inside. The past and the future are interlinked by experience. In considering events in the future, we often use past experiences to project possible future outcomes. This is both supportive and limiting, depending on what your beliefs are about this particular potential future experience. But again, the future only resides as an experience inside of us. It does not exist outside of ourselves.

The only real experience in time is *now*, an ever-changing present that is rushing past us, and the rest of our experiences in time are then only accessed inside of us.

1.1 Where and what is time?

If the question is posed about where time is in three-dimensional space, the response would have to be that time is everywhere. There is no place that time is not; from the smallest parts of the atom to the largest galaxies in the known universe, time is there. All of this matter we have just referenced, is ***in*** time. But what is this time that is everywhere, including inside of us?

We understand time as a measure of change, a difference between two moments in time. We take the length of time it takes for the Earth to spin once around its axis, and we call it a day. We take the length of time it takes the Earth to revolve around the sun and call it a year. We break these changes in rotation into smaller, more useful units. Instead of using fractions of a year, we have months, days, hours, minutes, and seconds. Our concept of time is closely related to spin. Even atomic clocks are based on the spin of atomic particles, measured in frequency or rotation.

We speak of our place in time, but we think of it as a spot in a long line, which stretches from the past, into the present, and on to the future. We speak of going forward or backward in time, but in reality, time is constantly increasing.

But could there be more to time? Is our place in time much more than a spot on a line? Could our place in time be similar to our place in three-dimensional space? This means that time would have to be multidimensional, likely three dimensions.

In this book, the objective is to ask questions of our understanding of time with regard to physics, and our own human consciousness. In this journey, hopefully a deeper understanding of time will emerge.

1.2 Which way must time go?

Physics treats time symmetrically. There is no reason that positive time [forward in time] cannot be replaced with negative time [backward in time].

Physics students, when studying electrodynamics, learn that Maxwell's equations do not limit electromagnetic waves from going forward in time only. Waves that move backward in time are solutions as well. Maxwell's equations apply to waves that go from the present into the future, as well as waves coming from the future to the present. These latter waves are known as time-reverse waves.

Since we only experience the time forward waves, the time-reverse waves are discarded as not physical solutions. But anyone interested in time is not immediately satisfied with discarding this potential solution. But this backward in time solution is not easily dealt with, so, for now, physics discards it.

We understand there is a limit on the experience we have of time. We see eggs break, which is an example of the general movement of events from order to disorder, but we do not see eggs reassemble after breaking, or in general, the universe of matter, spontaneously moving events from disorder to order.

This is a puzzle. Certainly, this follows our understanding of cause and effect. If this reverses, we need to come up with a different understanding of cause and effect that is not presently obvious to us. But this experience of the asymmetry of time, that is time only going forward, is different to what the physics equations say about the symmetry of time, which does not put any limitations on the direction of time, as mentioned previously.

But even given the overwhelming asymmetry in the forward direction of time, there are some phenomena and theories that allow for a reversing of time. A few of these will be discussed in the next two sections, so that a time-reverse situation can be visualized by the reader. Note, reverse and backward are not the same thing. A time-reverse wave can be generated while time moves forward. A time-backward wave is moving from the present time into the past, or from a future time into the present.

1.3 Time-reverse waves

There is an interesting technique in physics, where holograms are written in such a way that they create a very unusual mirror. The technical term for this holographic mirror is an Optical Phase Conjugation mirror—a mouthful. The important part is that it can generate a wave known as a time-reverse wave. Let's explore what this means, because, this time *reverse* wave is a good way to visualize a wave that is going *backward* in time, and visualizing time better is the objective here.

In a normal mirror that we use when getting ready in the morning, light comes from many directions to our face as we look into this mirror. This same light bounces from our face in many different directions as well. This allows you to see different parts of yourself in the mirror. If this did not happen, you could not see the different parts of your own body, never mind your face. Also, you would not be able to see anyone else in the mirror.

Now, in this unusual holographic mirror, if we have the same situation of standing in front of a mirror in the morning, the light bouncing off our face would go back in exactly, I mean exactly, the same path that it reached your face. Every photon of light coming from the light source will reflect off the holographic mirror and retract its path back to the exact same spot it came from in the light source. That means, if it came from the bottom left of the light, it goes to exactly that same place.

Now what happens if the regular mirror you use is replaced with a holographic mirror? Part of the light reflecting off your face from the lightbulbs will reflect toward the mirror, as before. But instead of continuing in different directions, as it does when reflecting off the normal mirror, all the photons of light go back and retrace their path to the exact same parts of your face they came from. The only light you could see in this holographic mirror is light that scattered from your eyeball. So, not a very practical mirror to use when getting ready in the morning, but these same unusual properties of this holographic mirror have other benefits.

How does this holographic mirror work? If we took a negative of an image of this zebra, shown in the figure on the next page on the left, illuminated it from behind, and then put in something that would distort this image, the viewer would see a *distorted image* as shown on the right. The object used to distort this image is a piece of glass like that of a shower door. [1]

If you place a regular mirror on the far side of the distorted glass, the reflected image goes back through distorted glass, and as expected, it becomes twice as distorted. This is shown as the *doubly-distorted image* on the bottom left in the image on the next page.

Now, if you replace the regular mirror with this holographic mirror (*phase conjugate mirror*), so that the same good image is passed twice through the same distorted glass, the final image seen on the left is the same as the *original image*. The distortion is now removed. How is this possible?

The holographic mirror generates what is called a time-reverse wave. All the distortions of the *distorted image* are reflected back through the distorted glass, but in a manner that is time-reversed.

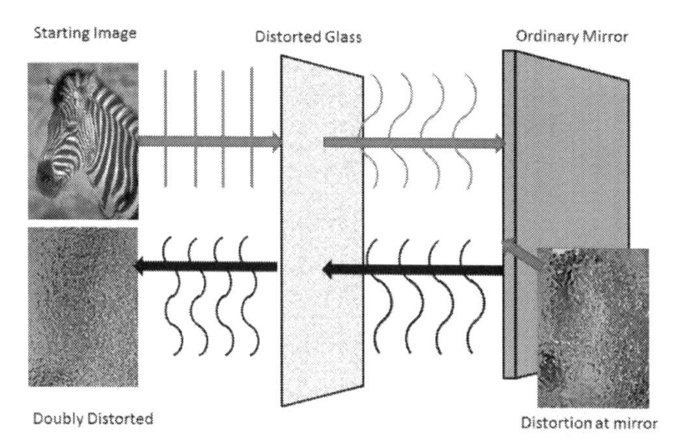

Ordinary Mirror

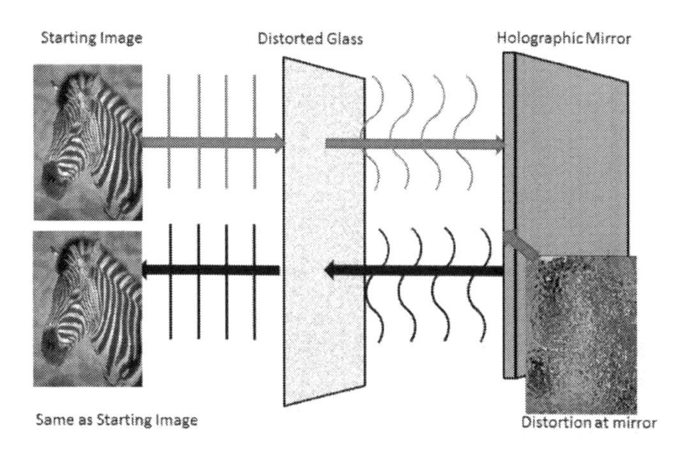

Time-Reverse Holographic Mirror

The time-reverse wave has all the distortion introduced on the first pass through the distorted glass, but in a time-reversed format. As this image information goes back through the distorted glass, these initial distortions are exactly canceled out by the distortions on the return pass through the distorted glass. The net effect is that all distortions are removed.

To understand how this is possible, it is necessary to dig into time-reverse waves a bit deeper by using an analogy of a marching band. In this analogy, the ten neatly organized rows of twelve band members are walking down the field. If an obstacle, like a set of stairs going up and down, were placed so only the middle four members of the marching band are affected, then this distortion is seen after the stairs, as the middle members would now be behind the musicians to either side of them due to the extra distance, and therefore the time, these four members traveled to go up and down the stairs.

To represent a reflection off a regular mirror, every row of the marching band has to walk to the end of the field and then turn around to walk back to the starting point. After the return to the starting point, the middle members are now twice as far behind, since, on their return leg, they had to go up and down these same stairs again. This represents the *doubly-distorted image* of the zebra seen in the figure on the previous page.

But with the holographic mirror, because it is three-dimensional, once the first row of marching band members reaches the end of the field, all the rows of marching band members turn around at the *same time*. Now the members who were behind due to the stairs, are now ahead. They are now in Time Reverse format. Those who were behind are ahead, those ahead are behind. As the marching band moves back to the starting point, the middle members go up and down the stairs again. But as they do so, they lose their lead by exactly the same time it takes to go up and down the stairs the first time.

Now, after the stairs, everyone is perfectly lined up again. So the holographic mirror that generates this time-reversed wave upon reflection causes the object that created the original distortion to '*undistort*' the original distortion. The result is that the time-reverse image now looks identical to the initial image. A very unusual mirror, indeed, one that allows order to be recreated out of disorder.

This understanding of this time-reverse phenomena will be useful in our exploration of time.

1.4 Matter and Antimatter

In the Paul Davies book *About Time*, he talks about anomalies in the symmetry of past and future time. He talks about how, for his thesis, he had to wrestle with the same issue regarding Maxwell's equations, the time forward and time backward solutions, mentioned previously. This caused him to ask all sorts of questions about how science understands time.

In his book, he describes nicely the issue of matter and antimatter, a theory developed by Paul Dirac. He states Dirac's motivation as follows:

Dirac wanted to know how a quantum particle such as an electron would behave when moving close to the speed of light. He discovered an equation that seemed to fit the bill, but he was baffled that every solution of the equation that described an electron came paired with a sort of mirror solution that did not seem to correspond to any known particle. After a certain amount of head-scratching, Dirac came up with a bold hypothesis. The "mirror" solutions, he claimed, correspond to particles that are identical to electrons, except their properties are reversed.

As Davies writes, Dirac called these particles with the reverse properties of electrons, positrons. Shortly after, positrons were detected in cosmic ray showers. Physicists came to call these 'mirror' particles, antimatter. For every particle, there is an antiparticle with reverse properties. Richard Feynman took this idea further and stated that the universe is full of electrons, labeled them as time-forward electrons, and positrons, which he labeled as time-reverse electrons.

Quoting Davies again:

Positrons are created twinned with electrons in violent encounters between gamma rays and matter. Typically, a gamma-ray photon colliding with an atom produces an electron-positron pair. The newborn electron flies off to enjoy a more or less permanent existence, but the poor positron faces hazards from the outset. If a positron runs into an electron (and the universe is packed with them), the pair will instantly annihilate, reversing the pair-creation process and giving back photons. This generally makes for a short career for the positron.

This explains why we measure all these time-forward electrons in our universe, but not many time-reverse electrons.

Further along, Davies states that physicist Feynman postulated that these electrons and positrons are the same particle, alternating between time-forward and time-reverse states. The conclusion would be that the universe is composed of half matter and half antimatter, but experiments show otherwise. There is more matter than antimatter, meaning that the universe leans in favor of time going forward.

Could this theory explain why we only measure time going forward? But where is all this time-reverse matter, or antimatter, which readily disappears from our time-forward universe? So far, there is no explanation.

This theory of matter and antimatter models two pieces of matter, our universe in which time goes forward, and another universe in which time is reversed. These time-forward particles of matter and time-reverse particles of antimatter unite into a photon of light.

This theory could accommodate new insights into the paradoxes of time. Let's continue on in the journey of time. [2]

1.5 Living in the past.

The previous sections dealt with time from a physics point of view. But it is important to look at time from our perspective, as conscious humans transiting through spacetime. One study really illuminated to me this fundamental relationship that we, as humans, have with time.

Two neurophysiologists, Benjamin Libet and Bertram Feinstein, did experiments to find out how humans respond to stimuli by measuring the time it took for a touch stimulus to reach the brain of a patient as an electrical signal, the time it took for this patient to respond by pressing a button, as well as the time it took for the patient to consciously articulate vocally an awareness of the stimulus.

The researchers reported it took one ten-thousandth of a second for the brain to respond to the electrical signal generated by the touch stimulus, one-tenth of a second for a person to press the button, and half a second for them to consciously respond that they felt this touch stimulus. In addition, none of the patients were aware they had pressed the button in response to the stimulus a full fourth-tenths of a second before they consciously recognized the event. [3]

In a sense, we are experiencing life in the past. By the time we respond consciously, the present moment has already come and gone, and there is a new, or potentially thousands of new, present moments in time that have happened and been recognized by our brain that we do not consciously know about yet.

A half a second does not seem like much from a human point of view, but from the perspective of an atom, billions or trillions of events on an atomic scale can happen in that time. A bit of a disturbing understanding.

From the experiment, we find out that our conscious response is at least 4,000 times slower than the brain's response. This suggests there's plenty of time for our brains to process the information and filter what we've just sensed and then report consciously preprocessed information. What is more alarming is that we're not aware of this preprocessing. There have been a lot of studies showing that our beliefs impact what we believe we see or hear.

The brain processes information from outside events through our senses, and from signals sent to the brain from cells. Information from both sources go through the unconscious part of the brain, which reacts before we even think of making a conscious response. Luckily, this saves us from all sorts of injuries to limbs, eyes, and other body parts. But this same preprocessing also colors the way we perceive information. Fortunately, with training, the neocortex part of the brain can intercept these responses before we react or express them, and with enough training, the unconscious part of our brain can be reprogrammed.

What is clear, though, is that we, as humans, consciously live in the past.

2 Quantum Theory

If you can fathom quantum mechanics
without getting dizzy, you don't get it.

Neils Bohr

2.1 Local versus nonlocal

Quantum Theory gave us the concepts that matter can be represented as a particle or a wave. The photon, a particle of light, can be represented either as a wave or a particle. But so can the electron, or any particle in physics.

To accurately describe an electron, as one example of a quantum particle, the best that can be done is that the electron can be described as an electron cloud, where the probability of finding it at any one location can be assigned. Quantum theory is a probability theory that has elegantly taken the obscure nature of matter represented as waves and combined this with their properties as particles. When an event occurs, all the probable places that the electron can be in the electron cloud is actualized into a particle at one location.

The famous dual-slit experiment in physics, normally done with either photons or electrons, demonstrates that these particles, in fact, do behave as a wave when not observed, but as a particle when observed. So matter behaves differently when observed. In this experiment, the Quantum observer effect is the act of taking measurements during the experiment.

In this chapter, we will come to learn that how we look at matter determines how it behaves. This can be tested in experiments with small atomic particles, but not with larger objects like cars, houses, trees, and the objects that make up our world. But every one of these larger objects is made up of these particles that do respond to whether we measure the properties of these atomic particles or not.

In the particle form, the particle can be treated as a separate object. It interacts with other objects as a particle that can be modeled as if it were a tiny spherical ball. Any interaction between the particle and the universe is through contact. This interaction is labeled as local, and all speeds of information associated with this interaction are at, or slower than, the speed of light.

But in wave form, these same objects behave in a completely different way. When the wave nature associated with two particles interact, patterns of light and dark bands occur. In this wave nature, or waveform, the particles interact with the universe in a manner that does not require contact, or even the forward progression of time. Distance is not crucial. These types of interactions are labeled as nonlocal because contact is not required. As we will come to learn, objects in wave form are linked with other objects or events throughout spacetime, allowing for information to travel at speeds much faster than the speed of light.

2.2 Who is watching whom?

In the dual-slit experiment, two beams of electrons, or photons, are directed toward a plate that has two narrow slits. When no measurements of these electrons are made, the pattern on a screen behind the plate with these two narrow slits represents exactly what you would expect from waves that move through narrow slits. The pattern has bands of light and dark areas, as shown below.

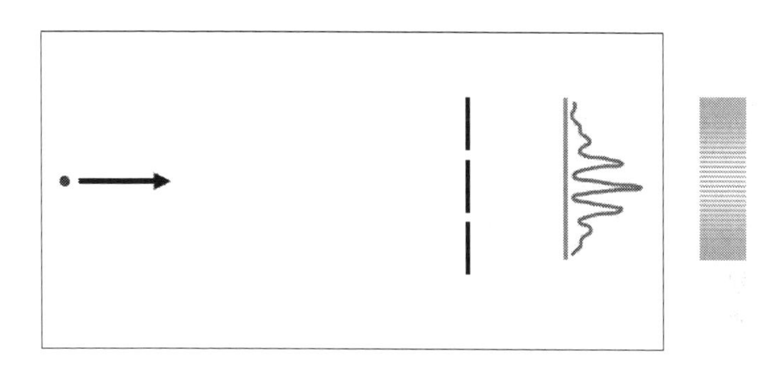

If you set up the beam so that the rate of electrons, or photons, or any other particle, coming from the source is so slow that a single electron can only go through either slit, you still see the wave pattern, even though there is no other wave for this electron to interact with. The electron interferes with itself.

If you block one of the slits, or you set up a measuring device, shown as an eye in the next figures, you now are attempting to observe the exact path the electron is going to take, and the pattern will change to one that is representative of particles shot through a narrow slit.

The detector behind the screen picks up the electron as a *particle* that goes through one slit or the other. As you would expect, most of the electrons go right through the slit to the detector, and fewer and fewer electrons are measured off axis.

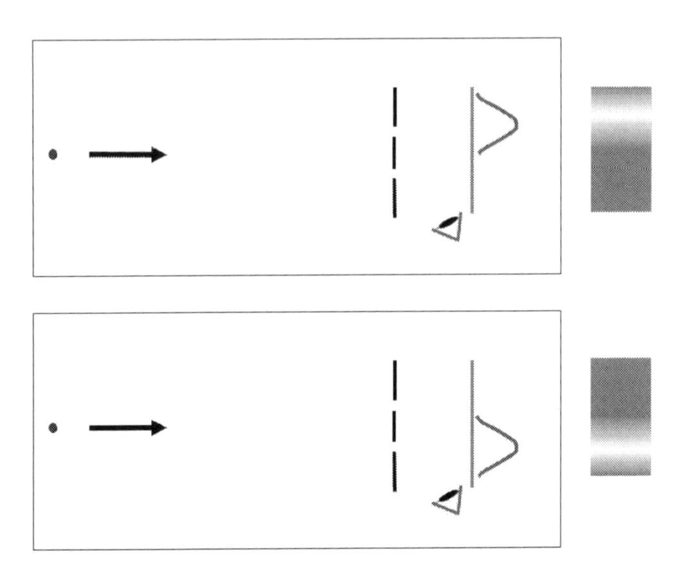

So who is watching whom? When we do not try to measure properties of the electrons, or photons, or any particles of matter, they are behaving as a wave. When we do try to measure their properties, they behave as a particle.

How does this interaction of observation work? When dealing with small quantum particles, the act of measuring the electron's position requires a method that involves photons. These photons have momentum. When the electron interfaces with these measuring photons, the electron scatters the photon. The direction of the scattering reveals the electron's position. In this measurement act, the scattering has imparted momentum from the photon to the electron, changing it. In the quantum world, you cannot observe what is going on without affecting it.

2.3 Again, now with time!

Now let's take this a step further. The observation process, when turned on, takes place after the electrons have already passed through the slit, but before they hit the detector behind the screen. The measurement process is designed to determine whether the electron went through the lower or the upper slit. The detection process is therefore taking place *after* the electrons have already gone through the slits and are on their way to the detector as a particle.

The experiment is set up so only one electron at a time is going through the slits. If the electron has gone through the slits and is on its way to the detector plate behind the slits, Quantum Theory tells us that without observation, the electron will behave as a wave.

Now, if the detector is turned on before the electron has hit the detector plate behind the slits, the observer process is now activated.

The electron will change from behaving as a wave to behaving as a particle. This will change the pattern on the detector plate behind the slits, which has to change to one of electrons behaving as particles going through the slits, *even though the electron has already gone through the slits as a wave.*

So the electron either behaves as both a wave and a particle all at the same time, or time is not as we understand it, a variable that just marches forward in time.

Remember, as a wave, the electron behaves as if there is another electron wave going through the other slit at the same time, but it is only one electron. If you close the other slit, you now know which slit it went through, and it behaves like a particle. The electron as a wave can determine by itself that the other slit is open or closed and behaves accordingly. Feeling dizzy yet?

Quantum Theory holds that all these possibilities are present at the same time. Only one of the potential outcomes becomes real when we measure the event. At the exact time of measurement, all the possibilities collapse into one actualized event, the measured outcome. Before the electron is measured, and after it has passed through the slits, the potential to be a particle and a wave is still there. Normally, the detector plate behind the screen marks which potential has been actualized. Since the detector plate behind the screen *cannot* determine which slit the electron came through, it is not an observer, and so the electron behaves as a wave. Remember, a wave pattern typically requires two waves to interfere. In the Quantum world, the interference pattern of light and dark bands can happen with just one electron, so the image on the detector plate cannot be used to determine if the electron wave went through the top or the bottom, since both paths produce the identical results. The measurement device, when set up to determine which slit the electron goes through, is an observer. It therefore can change what potential outcome the electron actualizes before it is recorded by the detector behind the screen.

This 'delayed choice' or 'Quantum Eraser,' as this experiment has been called, challenges our understanding of matter.

Dean Radin, in his book *Entangled Minds,* references a quote made by physicists. The first one, by Bruce Rosenblum and Fred Kuttner at the University of California follows:

The measurement problem arising from the quantum experiment does not necessarily imply that something "from the mind of the observer" affects the external world. The measurement problem does, however, hint that there is more to say about the physical than quantum theory says.

But a second quote by Pascaual Jordan gives a more direct statement:

Observations not only disturb what has to be measured, they produce it . . . We compel [the electron] to assume a definite position . . . We ourselves produce the results of the measurement.

2.4 Bell's non-local universe

Not every physicist was comfortable with the then relatively new Quantum Theory's postulate that all these potentials exist in a sort of quantum chaos before the measurement, at which point only one potential manifests into an actual reality. There were dissidents, one of whom proposed a new view of Quantum Theory.

Instead of matter having an either wave or particle nature, physicist David Bohm proposed that it has both all the time, not just in unmanifested reality, but in manifested reality as well. Matter, for example the electron, exists both as a particle and also as a pilot wave. This pilot wave interacts with reality in a distributed manner, giving feedback to the electron as a particle about how to behave with respect to position and momentum. This idea was embraced by other dissident physicists, but it presented its own problems. In order for this to work, the pilot wave has to interact with all other particles in the universe to determine the exact state of the universe. The universe then has to communicate the information regarding this state instantaneously to the particle. This requires information to be communicated at a speed much faster than the speed of light, violating Einstein's limit on the speed of light.

Another one of these dissident physicists, Einstein himself, was uneasy with the fact that this pilot wave would interact with all other particles instantaneously, no matter the distance. This meant that the information between the particles has to travel at much faster speeds than that of light (superluminal). Einstein, along with two other physicists, Podolsky and Rosen, devised a thought experiment called the EPR paradox.

In this paradox, two particles are generated in the same place and time and share a property that must stay the same for these two particles, a property called spin. If one particle changes its spin, the other must instantaneously change in order not to violate the laws of nature. This must be true even if the particles move in opposite directions away from each other at the speed of light, preventing any communication between them because the distance was too large to support the instantaneous change. The hope was that this paradox would highlight the impossibility of what he called the 'spooky action at a distance' aspect of Quantum Theory.

The EPR paradox was not solved in Einstein's lifetime, but when it was solved by John Bell's Theorem, it produced the exact result Einstein had hoped to dispel.

John Bell based his theorem on the pilot wave theories proposed by the dissident David Bohm. In his theorem, he states that the nature of reality is nonlocal. What does that mean?

One of the properties of Quantum particles that must be conserved is spin. If two particles are generated as 'entangled,' then their combined spin must always equal zero. Always, even if the distance between them is the opposite ends of the universe, if the spin of one particle changes—for example, from up to down—the other particle must instantaneously change its spin from down to up to compensate for this change. As mentioned previously, any information about the change in spin must be communicated at much higher speed than that of light. This implies that nature is responding to other parts of nature that this particle is never in direct contact with. This interaction without direct contact is the definition of *nonlocal*.

Nick Herbert, in his book *Quantum Reality*, defines nonlocal influences, with the qualifier 'if they exist,' as follows:

1. *Would not be mediated by fields or anything else. When A connects to B nonlocally, nothing crosses the intervening space, hence no amount of interposed matter can shield this interaction.*
2. *Does not diminish with distance. They are as potent at a million miles as at a millimeter.*
3. *Act instantaneously. The speed of their transmission is not limited by the velocity of light.*

In short, a nonlocal interaction is unmediated, unmitigated, and immediate. The immediate part requires information to travel much faster than the speed of light when the distances become astronomical. This speed has to be true if the laws of nature are not to be violated.

This property of entanglement has been verified now in many experiments around the world, and not only with quantum particles, but with atoms and molecules as well. Nonlocality is a part of nature and will help us to develop new insights into nature, time, and ourselves.

To complete the picture of a nonlocal universe, Dean Radin, in his book *Entangled Minds,* captures a quote I like from Robert Nadeau and Menas Kafatos, from their book *The Non-Local Universe:*

Also consider . . . that quantum entanglement grows exponentially with the number of particles involved in the original quantum state, and that there is no theoretical limit on the number of these entangled particles. If this is the case, the universe on a very basic level could be a vast web of particles, which remain in contact with another over any distance in 'no time' in the absence of the transfer of energy or information. This suggests, however strange or bizarre it might seem, that all of physical reality is a single quantum system that responds together to further interactions.

Matter at the most fundamental level can be modeled as behaving as if each particle is in contact with every other particle. The description of a web of particles remaining in contact with each other over any distance and in no time is a very good description of a single quantum system. Electrons, when cooled to a low enough temperature, behave as a single quantum system. Think of a huge flock of starlings, or fish, like sardines, that move almost instantaneously together in an orchestrated whole, as if all in the system are guided by a collective consciousness.

3 Zero-Point Energy

No point is more central that this,

that empty space is not empty.

John Wheeler

In this universe, there is a potential source of energy so large, that if it manifested into kinetic energy, that is, the energy of motion, it could vaporize all the matter in the material universe. Fortunately, this energy in the Zero-Point Field is perfectly uniform, so in this balanced state, only a tiny fraction of this Zero-Point Energy interfaces with matter.

Zero-Point Energy is a result of theoretical work done by Max Plank, the physicist who first quantized energy and started the whole Quantum Physics revolution.

This revolution started in an odd way. Max was a student in a graduate physics program in Germany, but was also a musician. He came from a musical family, and they were keen on him continuing as a musician, not as a scientist. He spoke with his professor, who told him to follow music, since most of the important ideas in physics had already been discovered and there was more opportunity in music. It seems the idea that physics had captured the most important ideas of the time was not uncommon at the end of the 19th century.

Luckily, Max did not heed his professor's advice and decided to continue his studies in physics, receiving his PhD, and eventually finding a post as a professor. He took up the problem of the issue of black-body radiation, which was in line with his current studies of energy and temperature. At that time, the classical theory of physics was unable to model accurately the experimental data measured for black-body radiation. Black-body radiation is the range of wavelengths emitted from a cavity in a blackbody, a material that is optically opaque, when this blackbody is heated to temperatures of thousands of degrees Celsius. Max was confident it would be solved soon, in accordance with the current understanding of physics, and set out to do so. But he was not successful with the current theories and had to take a new approach.

In 1901 Max Plank published his first paper on the radiation of a blackbody. Up to that point, the equations either were inaccurate at short wavelengths or at long wavelengths. His new equation was accurate across the whole range of wavelengths.

In order to make the equation work, Max had to introduce a new constant, h, which came to be known as Plank's Constant. What was disturbing about this new constant at the time is that the radiation had to be quantized. Energy was no longer continuous, but came in little packets called quanta. A quanta of energy is extremely small, but the implications to scientists at the time was that their idea of energy as a continuous phenomenon could be wrong. Physicists did not take this lightly, and a lively debate ensued. Even Max Plank himself was disturbed by the need for this constant h, the quanta of action as he called it, and tried in vain to get rid of it because of its implications. At the macro level, the level where classical physics predicts the behavior of matter, this has little impact. But at the microscopic level, specifically the level of atoms and particles, it has a radical impact. The theories of classical physics failed here and now; this new theory predicted that energy could only change in discrete amounts. The electron, for instance, had to jump from one energy level to another. No energy between these two energy levels could exist for the electron.

Max Plank worked on the blackbody issue for another ten years and then published a new paper in 1911. In this second paper, he had two terms in his equation. The first term had energy dependent on temperature, as expected, but in the second term, energy was not dependent on temperature, but only on frequency. Energy could now be stated as:

Energy = Plank's Constant x frequency

His constant, h, is necessary to convert frequency into energy.

Up to that time, in physics equations, the amount of energy in matter was dependent on the temperature of the material. At zero degrees Kelvin, a temperature of -273.15 degrees Celsius, matter stopped moving. With no matter moving, there is no vibration and, therefore, no energy. Since the second term was not dependent on temperature, it meant that there is energy in matter even at a temperature of zero degrees Kelvin. This second term came to be called Zero-Point Energy.

A new theory to model Quantum behavior was started with Plank's second equation. This theory, Stochastic Electrodynamics, uses a classical physics approach of electromagnetism with the 'jitter motion' of the Zero-Point Energy energizing matter. This jitter motion is also responsible for the uncertainty inherent in quantum measurements.

3.1 *Energy density of the Zero-Point Field.*

Zero-Point Energy, like the energy of electric and magnetic fields, is associated with a field appropriately called the Zero-Point Field. This Zero-Point Field is more fundamental than the Zero-Point Energy since the energy comes from the field. The Zero-Point Field has potential energy, that is, energy waiting to manifest. The Zero-Point Energy is a portion of this potential converted into kinetic energy, the energy of motion.

The concern with this second term of Max Plank's equation was that although the energy generated by each individual frequency is extremely small, the universe is expected to support an infinite number of frequencies, so the amount of energy from all these frequencies is infinite. The universe has infinite energy available, all the time. The fact that this amounted to an infinite amount of energy caused a lively debate among physicists about the validity of this Zero-Point Energy.

To deal with this infinite energy problem, a limit on the frequencies was set at the point where the fabric of space breaks down, a distance known as Plank's length. This Plank's length can be converted into Plank's frequency, and if you put this frequency limit, which is on the order of 10^{43} Hz, into Plank's second equation, the amount of energy density calculated is still extremely huge, on the order of 10^{107} joules/cm^3 (10^{95} g/cm^3). But it is no longer infinite.

Atomic energy, of which only a fraction is used to create the destructive force of atomic bombs, has an energy density of only 10^{37} joules/cm^3 (10^{14} g/cm^3). That gives a perspective on how much energy is stored up in the Zero-Point Field. As mentioned at the beginning of this chapter, the energy density of the Zero-Point Field is so large, it has the capacity to vaporize all the known matter in the universe. As mentioned in the first section, the reason that the Zero-Point Energy does not destroy matter is that it is perfectly uniform, so it cannot generate any forces to create this destruction.

The common example of how uniformity does not create any force is that the pressure of the atmosphere at sea level is 14 $lbs/inch^2$ but we do not feel this pressure because it is uniform inside and outside us.

For an analogy, let's assume you took a box and closed and sealed it at sea level. There is no pressure difference inside or outside the box. But if the box is taken to the depths of the ocean, it will be crushed by the outside force generated by the imbalance of pressures. The inside is still at the atmospheric pressure of sea level, but the outside of the box is at the tremendous pressures of thousands of feet of ocean depth.

Likewise, if the box is brought into the upper part of the atmosphere, it will explode because the force generated inside the box due to the sealed-in atmospheric pressure is much more than the pressure of the thin atmosphere thousands of feet above the surface of the earth.

The uniformity of the Zero-Point Energy is critical for the survival of the universe, as well as us. This extreme uniformity is also the reason it is undetectable. But if there is some imbalance, it then becomes detectable as motion. This motion, due to Zero-Point Energy coming out of the Zero-Point Field, was originally known by the German word '*zitterbewegung*,' or the English translation, jitter motion.

The jitter motion is associated with the particle form, and in this form, virtual photons come out of the Zero-Point Field for very short times and interface with matter. This virtual photon to matter interaction can generate a force, as will be shown later in this chapter.

This force, when considered in the waveform of nature, is dependent on the frequencies of these virtual photons.

This extremely high-energy density of the Zero-Point Field is much more noticeable at very small distances than at large distances. This is due to the fact that the energy in the Zero-Point Energy scales by the cube of the frequency.

Zero-Point Energy is proportional to frequency^3

The energy density increases significantly as the frequency gets higher. If the frequency is 100 times higher, the energy density of the Zero-Point Field is one million times higher. In a section below, we will see how this large energy density at higher frequencies, that is smaller wavelengths, validates the Zero-Point Field. A more technical approach that I have written about Three-Dimensional Time and Zero Point Energy can be found in Appendix A3.

3.2 Quantum Vacuum

There is another name that is also used to describe the Zero-Point Field, and that is the Quantum Vacuum. Since some of the scientists I reference call this field the Quantum Vacuum, it is important to clarify that the Zero-Point Energy comes from a field called by two names, the Zero-Point Field or the Quantum Vacuum. The reason for the two names is that the theoretical approach used to model the interaction between this field and matter is different. In this book, the approach is based on the Zero-Point Field, which uses the Stochastic Electrodynamic theory.

3.3 Mystery of the electron

One of the mysteries is how the electron, spinning around the atomic nucleus for billions of years at the speed of light, never runs out of energy. We hear predictions that in so many billion years, our planet will slowly spiral into our sun. But why does this fate not happen to the electrons?

In the theory of the Zero-Point Field, a place where the energy density is extremely high, Dr. Hal Putoff describes a mechanism of how the electron avoids this fate of crashing into the atomic nucleus because its 'tank' of energy is constantly refilled. The electron interacts with this Zero-Point Field and receives just enough energy to prevent this slow death spiral into the nucleus. [1][2]

If this energy transfer process is not in balance and too much energy is transferred, then the electron continues to gain energy, and it will overcome the electrostatic pull of the nucleus and free itself from the atom. If it does not get enough energy, it cannot overcome the electrostatic pull of the nucleus and would start this slow death spiral into the nucleus. So nature, with its usual propensity for balance, has organized that just the right amount of energy is transferred over billions of years to each and every electron in the universe to prevent this problem.

3.4 Validation – The Casimir Effect

This idea of the Zero-Point Field and Zero-Point Energy is very interesting, but can it be validated? Since the Zero-Point Field is undetectable, we have to focus on Zero-Point Energy, which is detectable through the energy of motion, this jitter motion, imparted to matter.

The experiment commonly referenced to validate the theory of Zero-Point Energy is the Casimir effect. The Casimir effect is measured experimentally with two plates that have a very tiny separation between them. The distance between the plates limits the number of frequencies, or more accurately, frequency modes, that are allowed between the plates. Since *all* of the allowable frequencies supported by the Zero Point Field exist outside, but only a subset of these frequencies can exist between these two plates, this difference manifests as a force that pushes the plates together.

An analogy might be a musical instrument, like a piano. Without any limitations, the piano supports all 88 frequencies of the keys. Assume all keys, that is piano strings, can be vibrated at the same time. We can add up the energy of all these piano strings at once. But let's assume that the piano is now limited to 22 keys, but not centered around middle c, but the 22 keys are at the high frequency end of the piano [shorter length strings]. The reason for this is that the Casimir plates would have the effect of limiting the vibrations of the piano to only higher frequencies as the plates move closer together. If we activate all 22 keys and add these energies, the expectation is that it would be clearly less than the 88 keys. But remember, the energy of the Zero-Point Field increases at higher frequencies and is dependent on the frequency cubed.

To give a perspective of how powerful this dependency of energy on the frequency cubed is, let's take an example of a single note. The lowest frequency supported by the piano is 27.5 Hertz, or vibrations per second, and the highest is 4,186.01 Hertz. If the energy difference of these two notes were just based on the frequency to the first power, the energy difference would be 152. But if the difference is based on frequency cubed, the energy difference between the lowest and highest piano key is approximately 3.5

million. So it is the high frequency, short piano strings, that would give this tremendous energy.

Because of this frequency cubed dependency, the energy difference between the 22 and 88 keys when they are played simultaneously is less than two percent. The 22 keys provide more than 98% of the energy density that is already achieved by playing all 88 keys. The rest of the 66 keys only provide less than two percent. This gives you an insight into why the high frequencies, like those at the atomic level, dominate the contribution to the energy density of the Zero-Point Energy value.

To measure a difference in the Casimir experiment, the distances between the plates has to be small enough to start to limit the number of high frequencies between the outside and the inside in order for it to become noticeable. This energy difference in the Casimir experiment manifests as a push because each frequency imparts a motion at the atomic level, which is small, but when it is added up over all the allowable frequencies supported by the Zero-Point Field, is a significant value.

It is easy to visualize that as the distance between the plates gets smaller and smaller, the difference in energies between inside and outside gets larger.

Experiments done with the Casimir effect have validated theoretical calculations for Zero-Point Energy to approximately five percent, a very good correlation.

Much speculation has been written about techniques to extract some of this Zero-Point Energy into usable energy. Clearly, there is a huge amount of energy stored up in the Zero-Point Field waiting to be tapped responsibly with the correct technique. If done responsibly, it would solve all our human energy requirements for as long as we can imagine. If done irresponsibly, we would only create more destructive technology, which would make nuclear energy look like child's play.

3.5 Zero-Point Field properties

The key properties of the Zero-Point Field are:

1. The extreme uniformity of the Zero-Point Field makes it undetectable.
2. The extremely high energy density of Zero-Point Field is potential energy, energy waiting to manifest.
3. When the uniformity of the Zero-Point Field is changed, this potential energy now manifests as kinetic energy, the energy of motion. This motion is known as jitter motion.
4. The energy density of the Zero-Point Field is dependent on frequency cubed, so high frequencies make the most difference.
5. The fact that motion is the interface between this Zero-Point Field and matter is very important, since the interface between space and time is motion.

4 Spacetime – current understanding.

I was born not knowing and have had only a little time to change that here and there.

Richard Feynman

Initially, the units of measurement for distance were defined by the human body. The Egyptian cubit was based on the length of the arm, from the elbow to the tip of the extended finger. The Roman mile was based on a 1,000 human paces, and the Roman foot, well, obviously based on a human foot. The recent addition of meters by the French is based on the distance from the North Pole to the equation divided by ten million.

In each case, a special rod was made to represent this dimension and was kept securely by the administrations of the time. Copies of this special rod were made to ensure that all the rulers at the time measured the same distance between two points in space.

The units of time were based on the cyclical positions of the sun and the moon in the heavens. To make these units more useful, the days were divided by 24 into hours, and then divided once by 60 into units of minutes and then again by 60 to get units of seconds.

Then, at the beginning of the 20th century, two scientists started a revolution that forever changed the way we could relate to space and time. The two scientists had a working relationship; one was Albert Einstein, and the other was Einstein's professor at the University, Hermann Minknowski.

Einstein's Special Theory of Relativity had length contract and time slow down when velocities are increased. These effects are not really noticeable until the velocity comes close to the speed of light. Since only atomic particles can come close to this velocity, this effect on space and time is only measurable for these small particles. At the speeds we experience in life, the Special Theory of Relativity results are no different in our everyday experience.

4.1 Moving in Three-Dimensional Space.

If three-dimensional space is considered without time, and you are in a specific location in space, you will stay there forever, since there is no way for change to happen.

In order for change to happen, like a change from one location to another, then a variable of time is required. How does time enter into this space relationship to create change? It manifests in speed, or velocity, which is change in location in space divided by change in time. This velocity is measured in units of space divided by units of time. In our car, this is commonly done in miles/hour or kilometer/hour.

Adding time into space allows *change*. So let's see how time continues to dance with this property we call change.

Now, with speed, movement is possible, and you can move from one location in space to another location in space within a specific '*amount*' of time. Many remember their initial exposure to time in physics being an exercise where an object is at some location at t_0, and then having to predict its behavior at t_1 based on some equation of motion.

But the real philosophical implications of speed is never investigated in these classes or experiments. The fact that speed is change is obvious, but that it is a ratio of change is less obvious, and will be explored in this book.

Speed is understood well and seems intuitive. Velocity is typically only used for science and engineering applications. Velocity is speed, but it also accounts for direction in space. Speed only needs one term, and it is defined as a magnitude, such as miles/hour of a car. It is known as a scalar variable. It can go up and down. When going up, it is positive, when going down it is negative. Pretty straightforward.

As mentioned, velocity has direction built in. If velocity is in three dimensions, it has three terms. Normally, three-dimensional space is defined with a direction in x, a direction in y and a direction in z. But to make this relevant, let's use an example of you sitting in a chair. If you want to get out of the chair, you need to move, so you need speed, or velocity. Let's assume the chair is situated by a fire in a wooded campsite. It is dark out, and suddenly, out of the darkness of the forest, you see a black bear coming toward you. Your reaction, measured in speed, is calculated by taking the distance you cover rapidly, exiting your comfortable chair and dividing it by the time it took you to cover that distance. We understand why the speed would be high given the situation, but speed does not tell us in which direction you are going. If this was the only parameter we could use, it would be important that your speed is faster than that of the bear. But that is not all you need to know. What direction did you take off in relative to the direction of the bear who is approaching you? If you headed toward the closest tree, what direction is the tree relative to the direction of the bear coming toward you? Velocity takes care of this.

In calculating your velocity, we would calculate your speed along the north-south axis, speed along the east-west axis, and speed along the up-down (height) axis. This last axis is probably crucial, since the speed at which you can climb up a tree might be the difference between a nasty encounter and one where you get to tell your friends about how close it came. Additionally, the speed of the bear along these three axes is crucial to determining ahead of time whether you have the opportunity to get out of the bear's way in time along these same three axes. Since velocity has a speed along each one of these axes, it is defined as a vector. Vectors have direction built into them.

Now, in this example, there is another term that is important, and that is acceleration. Acceleration is the change in velocity in a given 'amount' of time. So using time again *changes* velocity. This shows up in the units of accelerations, since acceleration is measured in units of space divided by units of time squared, or meters/second^2. This is the same as a double *change* in locations, since not only did you change your location, but you are rapidly changing the speed at which you are leaving that location, hence the *change*^{2} coming from time^2 term.

We assume the bear is moving toward you at a certain speed. Unfortunately, you notice him while you are enjoying a fire and your speed, or velocity, is zero. Now your ability to avoid an unfortunate encounter is based on your ability to change your velocity from zero to a speed as fast or faster than the bear in a very short amount of time. So you have two changes you need to make: The first one is location, and the second is your speed. That is acceleration, and it is also a vector because you are increasing your velocity in a particular direction—hopefully the right direction relative to the bear.

The force required to move you, your mass, by this necessary acceleration to stay away from the bear, is what you need to generate. This comes from Newton's second law, Force equals mass times acceleration.

From this example, the understanding is that in our current definition of spacetime, space is a vector since it has multiple axes, specifically three dimensions, but time is a scalar, since it can only increase or decrease and has no dependency on the direction of space. It is the same whether you move along the north-south, east-west, or up-down axis. At this point, you might say that the east-west axis does change time. When flying east we 'lose' time in our day, and when we fly west, we 'gain' time in our day. But what we are dealing with here is the structure we as humans have set up to mark the passage of time in our lives. The speed calculated in the Western Hemisphere is not different in the Eastern Hemisphere for the same distance. So this is what is meant that time is not dependent on direction of space.

4.2 1D Time in 3D Space

The properties of 1D Time have already been discussed in Chapter 1. What is important to stress is that 1D Time appears everywhere. To reiterate, there is no place in space where time is not. It is very uniform, for if it were not, we would have radically different effects based on different values of time. Luckily, we can use the same method to measure time in atomic events, human events, and in cosmic events because of its uniformity.

4.3 Relatively speaking

In physics, in order to measure an event or phenomenon, you have to pick a reference point.

Commonly, for humans, that is the surface of the earth. We take our reference point of the surface of the earth for granted, experiencing it as a location that is stationary. But it is far from stationary.

Any point on the surface of the earth is rotating with a velocity about the axis of the earth at anywhere from approximately 1,000 mph around the equator to 0 mph at the axis of rotation. Our planet's velocity around the sun is roughly 67 thousand mph, and our solar system's velocity around the Milky Way Galaxy is roughly half a million miles an hour. Hardly stationary.

Our sense that our world is stationary is because we live on the surface of a planet with a radius of about 4,000 miles. Typically, during our lives, we never experience a change from this radius of more than one hundredth of a percent when we fly in a plane, and a minute fraction of that when we go scuba diving. Our whole life is constrained to roughly 4,000 miles plus 0.01 percent and minus 0.000005 percent.

Luckily for us, nature is not sensitive to constant velocity, only to a change in velocity. The joke is, that it is not the falling that kills you, but the sudden change in velocity when you hit the ground. A change in velocity is acceleration, or deceleration when slowing down, like hitting the ground. It is when force manifests due to your sudden deceleration as your velocity suddenly goes to zero that you need to worry about when falling.

Even a tiny change in velocity of our planet's rotation, or orbit around the sun, would have catastrophic results for humans living on this planet, so the solar system and the universe have to be extremely well-behaved for us to thrive.

Velocity is very important, because where we are changes how we perceive things. Let's take the example of a train, one that has glass sides so the activities inside the train can be easily seen by those outside. Let's have a person on the train take a ball and bounce it off the train floor, catching it, and then throwing it back down in a constant routine.

For the person throwing the ball, and all those on the train, the ball leaves his hand, then bounces off the floor back to his hand. Any physical measurement of distance or velocity of this ball would describe it in this way as shown in the 'Inside Viewpoint' below.

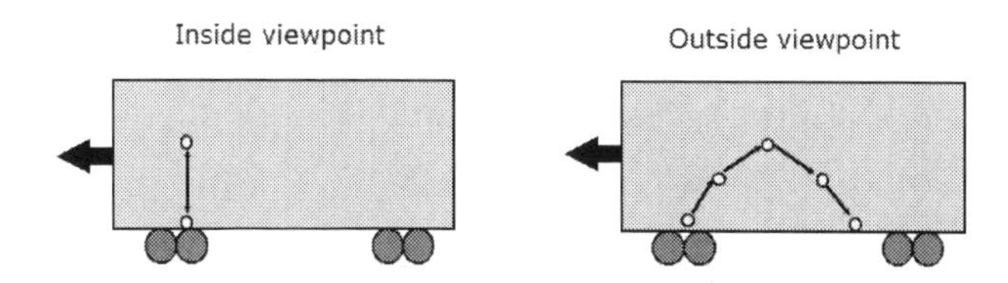

But if you were to witness this same activity of bouncing the ball outside the train while it was moving, you would see, and measure, a very different trajectory of the ball. The 'outside viewpoint' figure above shows how someone outside would see the ball moving.

Its velocity of the train has changed the perspective between the two groups. Now, philosophically, which view is correct? They both are; it just depends on your point of view, or your reference. In the inside viewpoint case, your viewpoint is changing along with the velocity of the train so that the velocity effect is canceled out. In the other, the observer watching you is at a different velocity, in this example, standing on the platform at the train station.

The difference in these two viewpoints changes the experience and measurements of these two people. But their viewpoints can be correlated by using velocity.

The Special Theory of Relativity takes away the notion of a common universal present moment. This effect is really pronounced when velocities approach the speed of light. For all humans on Earth, the speed of light is so fast that the largest difference in a present moment might differ by a little more than a tenth of a second at any point on the surface of the earth. From the experiment discussed in Chapter 1, we know we consciously will not be aware of changes on this time scale.

But in situations where the distance between travelers is huge, that is, where is takes years for light to travel between one traveler and another, the present moment is not the same. This can be explained in an attempt to synchronize multiple clocks in different locations. The fastest way that any measurements can be communicated is at the speed of light. The speed of light puts a limit on how fast measurements can be synchronized that are separated in space. For example, clocks on Earth, the sun, and Alpha Centauri need to be synchronized with a light pulse. The clock on Earth is synchronized immediately when the pulse is generated, the clock on the sun receives the light pulse a little over 8.3 minutes later, and the clock on Alpha Centauri 4.4 years later. People on Earth, the sun, and nearest star have different starting times, and therefore, will measure time differently. If everyone wants to measure the same event at a fixed distance, a delay can be factored in for each of the clocks, and also for the time it takes for the light from the distant event to reach each viewer. But all the distances have to be premeasured to factor in the right delays.

If the three people are moving in constantly changing directions, and at different velocities, the delays will change constantly, and there is no way to update them without having new delays already in the communication of the last delay. So the times experienced by people using these three clocks will be different, not only in this simple manner, but also in more complex ways.

4.4 Spacetime Intervals

Professor Minkowski took his students' idea and clarified them by combining space and time into a four-dimensional spacetime. Time was added as an equivalent fourth dimension to space.

Instead of using space and time directly, since these are different for different observers, a new variable was needed that would be the same for all observers. The variable called spacetime interval, which is made up of a component that is the separation in space and separation in time, was found to be consistent for all observers and events in spacetime.

Space and time now are combined into spacetime, which was a novel idea at the time. In spacetime, we now talk in terms of events and intervals. [1]

In order to combine time with space, either units of time are converted into units of space or vice versa. Typically, time is multiplied by the speed of light, so now, time is in units of distance [space]. This does not mean that time is the same as space, but that time is measured in equivalent units, that is, units of distance of light traveled in an interval of time. If we know how long it takes to travel to a friend's house and we know the speed we can travel at, then we can figure the distance. An hour's drive at an average of 60 mph gives us 60 miles. Time multiplied by velocity gives an *equivalent* distance for that time.

With the new form defined by Minkowski, the locations of events in spacetime could now be plotted as a four-dimensional vector, three dimensions of space, and one dimension of time.

If you could move at the speed of light, your reference frame moves along with you. In this reference frame, the passage of time slows down, since the increment of time changes. But the length of this four-dimensional line, called an extension, must stay the same, so the lengths in this moving reference frame contract. The increment of time changes in noticeable ways as the velocity of one person relative to the other starts to reach the speed of light (assuming for the sake of this thought experiment that a person can move that fast).

The Special Theory of Relativity and Minkowski spacetime diagrams change radically what we considered absolute space and absolute time. No longer could we expect how we measured the universe around us to be the same for all observers if they are moving at different velocities. Velocity is *the* variable that determines how the definitions of an increment in time and length change.

As we saw Minkowski do earlier, he coupled 3D Space to 1D Time via the speed of light. If you take distance and multiply it by velocity, the result is time.

Velocity is the coupling of space and time, and is defined as

Velocity = change in space[distance] / change in time

The speed of light, a constant, can be used to convert between space units and time units. Just like the reference rods that defined the units of cubits, feet, and meters for the different civilizations, the speed of light is a reference between these two reference frames, one in space and one in time.

But the speed of light is more than a conversion factor, it is the *ratio* of space *to* time and time *to* space. This ratio of space to time is the basis of Dewey Larson's Reciprocal System Theory.

4.5 Energy and velocity

Now, the interesting feature of energy is that it has no direction in space. It is always a scalar, a one-dimensional variable, that increases or decreases, but never has a preferential direction. This is important to remember when energy in time is discussed.

If energy is dependent on a variable that has a direction, like velocity, it must be dependent on the square of the variable, since a directional dependent variable [vector] times another directional dependent variable [vector] is a one-dimensional variable [scalar].

From a classical point of view, for a fixed mass, the velocity is the energy of the system. The equation is in the form of,

$$Energy = \frac{1}{2}\ mass \times velocity^2.$$

The amount of energy quickly changes with velocity. Double the speed you are going, and the amount of energy you have generated has increased by four. Triple the speed, and now the energy has increased by nine times.

Energy can also be related to mass and frequency. Einstein's famous equation,

$$Energy = mass \times speed\ of\ light^2$$

correlates energy to mass. The amount of energy in mass is 186,000 squared, or roughly 3.5 trillion times the amount of mass. An incredible amount of energy can be generated from mass.

Also from Quantum Theory, energy is directly related frequency, or inverse time, via Plank's constant.

Energy = Plank's constant x frequency.

Later, the frequency will be shown to be a special form of velocity.

In each case, energy is derived from velocity, or motion. In the first equation, it is easy to understand since velocity is motion. In the second, mass will be shown to be equivalent to inverse velocity, and the speed of light is clearly motion. In the third equation, frequency is equated to a velocity where the unit of space is equal to one. So, as Dewey Larson points out: ***motion is more fundamental than energy.***

4.6 Properties of 3D Space – 1D Time

It is beneficial at this point to define and clarify a number of the terms that will be used throughout this discussion and will be important when the topics of 3D Time are discussed.

To help clarify the properties of 3D Space and 1D Time [Vector Space and Scalar Time], a summary list will help.

1. Three-Dimensional Space allows the location and shape of an object to be defined.
2. The first property of space is location. Measured in unit's L for Length but will also be represented as S for Space.
3. Embedded in this first property is shape. Length is Space, Area is $Space^2$ and Volume is $Space^3$.
4. The second property of space is velocity. Measured in units of Space / Time or meters / second. Velocity in Space manifests in the change in location.

5. Velocity generates energy in space.
6. The third property of space is acceleration. Measured in units of Space / Time^2 or meters / second^2. Acceleration is the change in velocity, or the change^2 of location. Here, change manifests as Time^2.
7. Space is the local part of our natural world. Any exchange of energy or information is done with contact between parts of the universe in some form. Therefore, 3D Space has properties defined as 'local' and is associated with the particle-like nature of matter.
8. All motion in 3D Space is equal to or less than the speed of light.
9. 4D spacetime is three dimensions of space and one dimension of time. So 4D spacetime is defined mathematically as Vector Space and Scalar Time.
10. To change anything in 3D Space requires time. The most obvious example is velocity and acceleration, both changes of location. Time in Space manifests in the form of 1 / time, since velocity is space multiplied by 1 / time and acceleration is space multiplied by 1 / time^2.
11. Since Time, a one-dimensional variable or scalar, is everywhere in 3D Space and does not require contact and is therefore referred to as <u>nonlocal.</u> 1D Time is associated with the wavelike nature of matter.
12. Time, as a one-dimensional variable, can go up or down in magnitude, but has no direction. We speak of time going forward, but in reality, it's increasing in magnitude. Going back in time is reducing its magnitude.

5 Velocity – the ratio of Space To Time.

By far the most important consequence of the conceptual revolution brought about in physics by relativity and quantum theory lies not in such details as that meter sticks shorten when they move or that simultaneous position and momentum have no meaning, but in the insight that we had not been using our minds properly and that it is important to find out how to do so.

Percy W. Bridgman

5.1 A universe in constant motion

In the previous chapter, we discussed how we take for granted our reference point in this universe, the surface of the earth. We feel that the world around us, the ground, the trees, the mountains, houses, and buildings are still. We expect to find these in the same place they were yesterday, and the day before that. The rocks in our neighborhood and in the woods around us seem motionless. But this is just an illusion, since the world around us is far from stationary, as pointed out in the previous chapter.

Even how we see the world is based on motion. Our eyes are in constant motion. Without motion, the scene you look at would disappear. In a day, our eyes make about 200,000 movements up and down and side to side. We're not aware of this, but without it, our vision wouldn't work. We perceive things with our eyes because our eyes are moving or because the object is moving. In fact, our eyes are designed to see motion in preference to stationary objects. When technology is used by scientists to counteract the movements of the eyes, the images the person perceives fade away. The neural system of the eye requires constant stimulation due to motion.

The basic building parts of our material universe are moving at the speed of light. So there is no part of our material universe that is not moving. And motion involves both space and time, but not just these two in a random way. Velocity is the ratio of space *to* time, defining time's relationship to space. Let's explore what this means.

5.2 The most famous velocity – speed of light

We discussed how velocity can be used to convert space units into time units. The Special Theory of Relativity shows how central the velocity of your frame of reference is to how you measure time and space.

In so many equations in physics, this constant speed of light, represented by '*c*,' is one of the foundations of how we understand our universe. What do we understand about the speed of light?

1. Light travels 186,000 miles in one second.
2. That unit of distance represents about ¾ of the distance to the moon in one second. For most of our purposes, except for specific engineering

and science applications, it is a very large unit of distance and not very useful.

3. The unit of one second we can relate to. We can count it by vocalizing 1,001, 1,002, 1,003, etc, and it is easily measured by our watches. We use the second as a basis for tempo in music.
4. Since the distance covered in one second is so large, it might be easier to reduce the distance to a foot, a unit we can relate to. To keep the speed of light constant, we have to reduce time, in this case, to roughly a billionth of a second.
5. Now we can relate to the unit for distance, but not the unit for time. This time is better suited for atomic events. But a foot is far too large for atomic scales.

From a human perspective, the speed of light does not represent a useful frame of reference in both space and time simultaneously. As we currently define this constant, the time makes sense, but the distance is enormous. Readjust it and the distance makes sense, but the time is miniscule. What does that say about space and time when they are combined?

5.3 Speed and Velocity

Previously, the description of how speed does not have a direction but velocity does was discussed in the example of the bear and the person in the woods.

The speed of light has no direction, which is why it can be represented by '*c,*' a single variable, called a scalar. Photons of light, not just visible light, but any electromagnetic waves, have no mass and move in all directions equally, so photons have no preferential direction. This is not to say we cannot constrain photons by experiment to travel in a particular direction, and in doing so, can now represent these photons' speeds as a velocity.

5.4 Ratios

In this section, I want to discuss ratios that are commonly used in geometry and in physics. It helps to understand the implications of ratios and what the ratio of space *to* time means.

5.4.1 Pi and Phi

Our most common experience with a ratio is with Pi. We are all taught that there is a direct relationship between the radius of a circle and the circumference and area. This relationship is a fixed ratio called Pi.

Another ratio is Phi. Phi is a ratio of lengths, based on the Fibonacci series. This ratio expresses itself in the geometry of our own bodies, as well as in the natural world around us. The lengths of so many of our bones have a Phi ratio. For example, our arms, above and below our elbows, is the Phi ratio, the bones in our hands, the patterns of flowers, trees, fruits, as well as the spirals seen in the universe also correlate this ratio Phi.

5.4.2 Index of refraction

The experience of seeing a pole looking like it is bent when part of it is under water is a direct result of the ratio of the speed of light in vacuum to the speed of light in matter—in this case, water. This ratio of the two different speeds of light is called the Index of Refraction of the material. This same effect makes fish look like they are closer to the surface of the water than they actually are.

This ratio of the speed of light in different materials, all referenced to the speed of light in vacuum, is used to determine the amount of bending of the light in matter. The image of a prism with white light coming in on one side and a rainbow of color coming out of the prism is due to the refractive index.

Since the index of Refraction, represented by the variable *n,* is greater than one, this means that the speed of light decreases from the speed in a vacuum to a slower speed in matter, in this example, water. This could mean that either the amount of distance decreased, or the amount of time increased. Both have the effect of decreasing velocity.

5.5 Speed of light as a ratio of change

In the previous section, the idea of ratios was discussed to help the reader see velocity in a new light, as a ratio of space *to* time.

In physics, we learn to deal with velocity as a variable that represents a change in distance divided by a change in time.

In order to convert time into equivalent space units and vice versa, the speed of light is necessary. For example, time multiplied by velocity gives distance. So to include time in an equation that has units of distance, it is necessary to include not just t [time], but ct [speed of light multiplied by time]. The reverse is true. To include distance in an equation with units of time, distance has to be divided by the speed of light, so time equals distance / c.

The speed of light is used to convert time to space, and space to time. This is not new.

What *is* new in this Reciprocal System Theory, is that it is a *ratio*. It is the ratio of space, represented by distance, to time. What does this ratio tell us?

Let's look at the representation of velocity,

Velocity = change in distance / change in time

This can also be stated as:

Velocity = change in space / change in time

To increase velocity, we can increase the change in the distance covered for the same amount of time. We can also increase velocity by decreasing the amount of time to cover the same distance.

We can summarize this relationship between space and time as follows:

Increasing distance has the same effect on velocity as decreasing time.

The same is true for the inverse.

Decreasing distance has the same effect on velocity as increasing time.

That is

Space = 1 / Time

and

Time = 1 / Space

With regard to velocity, space and time are reciprocal. Increasing one is compatible with decreasing the other. So if space expands, time contracts, and as time expands, space contracts. Understanding this reciprocal relationship is crucial to understanding multidimensional space and multidimensional time.

Since the speed of light is a constant, if distance increases, the change in time has to increase to compensate for the change in distance. From Quantum Theory, we know that energy is quantized, and with that, space and time are quantized. So if the distance increases from one unit of distance to ten, time has to increase from one unit to ten units, since

1 unit of distance / 1 unit of time = speed of light

and so does

10 units of distance / 10 units of time = speed of light

The speed of light can be generalized as

Speed of light = m units of distance / n units of time

Where m units must be equal to n units. If m and n are not equal, then the speed is either slower [m < n] or faster than the speed of light [m > n].

There can be scenarios where m units for distance is equal to one. Frequency is an example. If m equals one, then the equation for velocity becomes 1 / time, the expression for frequency. *So frequency is a special form of velocity, which only involves changes in time.*

This also applies to wavelength, since c equals frequency times wavelength. So wavelength is a special form of velocity where n always equals 1, therefore wavelength only involves changes in space.

6 Properties of 3D Time and 1D Space.

Hide nothing; for time, which sees all and hears all, exposes all.

Sophocles.

In this chapter, time will be expanded from the current state as an additional dimension in spacetime into a four-dimensional reality, which is the reciprocal of the spacetime we understand.

In the Reciprocal System Theory, one of the postulates is that there exists another reality, complementing the spacetime reality of three-dimensional space and one-dimensional time we currently understand. This new reality has three-dimensional time and one-dimensional space. As stated before, the relationship between the two is the speed of light. This means that distance is equivalent to 1 / time and time equivalent to 1 / space.

Velocity is three-dimensional space / one-dimensional time. The inverse of velocity, one / velocity is three-dimensional time / one-dimensional space. These two four-dimensional domains are reciprocal of each other, but the relationship is a one-dimensional correlation since it is the speed of light, not the velocity of light, that connects the two. The connection can only be a ratio. [1]

I found it difficult to understand and visualize initially and did not find any explanation in the Reciprocal System Theory that helped me. What I will do in this chapter is take the concept of three-dimensional time and one-dimensional space that Dewey Larson has postulated and put them into terms that I found easier to understand, and hopefully you will as well. What I say here might be unacceptable to those who follow the Reciprocal System Theory, but my approach has helped give me new insights into space and time, and I hope it will do the same for you, the reader. I also apply some simple mathematics to help explain the derivation of some of Larson's key postulates. Lastly, I use a few of my own models to expand the understanding of multidimensional time.

In Dewey Larson's book, *Nothing But Motion*, he refers to 3D Space and 1D Time, the one we are familiar with, as timespace and calls it the material sector. He calls 3D Time and 1D Space as spacetime and calls that the cosmic sector, the reciprocal of what we are familiar with. This timespace and spacetime designation, I find, is unfortunate because it adds confusion to a concept that is already a challenge to describe. While I am tempted to reverse this designation, since spacetime to me, and I assume the rest of the readers, means 3D Space and 1D Time and timespace would then mean 3D Time and 1D Space, it would only add another layer of confusion between these two domains.

From now on, I will use 4D Space to designate both 3D Space and 1D Time and 4D Time to designate both 3D Time and 1D Space. The important part here is that 4D Space has motion, whereas 3D Space does not because there is no time. The same is true for 4D Time. Motion in 4D Time is possible because it has 1D Space, whereas 3D Time does not.

In mathematics, there is a technique that transforms a scalar, a one-dimensional variable that has no direction, and turns it into a multidimensional variable that has direction. In turn, there is also a technique that takes a multidimensional variable called a vector and transforms it into a one-dimensional scalar variable, stripping away the directional information. So there is a way to take 3D Space, or 3D Time, and turn it into 1D Space, or 1D Time. The reverse is true as well, in that 1D Space or 1D Time, can be transformed into 3D Space or 3D Time. These dimensional realities are coupled and can be transformed one to the other.

Even if you find this chapter challenging, continue on to the next, where 4D Space and 4D Time are considered together, and by rereading these two chapters, the concepts will become easier to visualize.

6.1 Three-Dimensional Time

If time is in fact three-dimensional, that means there are locations in time. We use language for this, saying 'I was in a different place in time,' but we are referring to our consciousness about a particular topic.

With 3D Time, we need to understand what the three axes of time might mean. In 3D Space, the first axis is length, the second is width, and the third is height. With length and width, we define the area of an object's shape, and with the third variable of height, we can define the volume of an object's shape.

In 3D Time, we naturally would describe one axis as a period of time, or length of time. The other two axes would be an area that intersects each moment along the direction of this first axis of time. The area of this plane would represent an extension of the present moment in time. But we have used two axes to capture the present moment, without being able to distinguish them. Maybe there is a better way to describe the properties of 3D Space and 3D Time than the length of the fundamental unit, distance, and period.

Let's look at it a different way. For 3D Space, we can define fundamental properties that are observable. Observable is the key parameter. Observable does not just mean visible by the eye, but that the property can be measured by instrumentation. Nonobservable means this property is invisible, but more importantly, we cannot measure it with any of the equipment we currently have. But, we can calculate Nonobservable properties with equations that have Observable properties in them. For instance, Energy equals ½ mass times velocity squared. We can observe velocity, and we can observe mass, but we cannot observe energy. We cannot see it, and the only way to measure it is to measure mass and velocity, and then calculate it. So Energy is nonobservable but is calculated through observable properties. Richard Feynman, a famous physicist, put it this way: '*I don't know what energy is, but if you have plenty of time, I can teach you how to calculate it.*'

The first property of 3D Space that comes to mind is location. We can observe objects as having a location in 3D Space. We also observe objects as having shape. We can also change our locations in space by moving in 3D Space, and as we now understand more clearly, we cannot change locations in space without time. When we move in space, we create energy in the form of movement. We can observe the motion, but we cannot observe the energy, so we can only list the motion as observable. When we change the rate we are moving at, we have acceleration. If the object has mass, acceleration results in a force. But the force is not observable, so we can only list the acceleration. To summarize these observable properties, we have:

1. *Observable Location in 3D Space*
2. *Observable Shape in 3D Space*
3. *Observable Motion in 3D Space*
4. *Observable Acceleration in 3D Space*

Now let's try the same exercise for time. We assume we are beings living *in* 3D Time and we can see in these time dimensions. With 3D Time, there is an observable location in time. With three axes, each object in 3D Time has an observable shape. To get from one location in time to another, motion is required. But what does motion in 3D Time mean? We will address that in the next section. The motion is observable, but is the energy? We know it is not for multidimensional space, but we do not have enough of an understanding to know if it is observable in multidimensional time.

To change the rate of motion, acceleration in 3D Time is necessary. What does acceleration in 3D Time mean? Again, the acceleration would be observable, but would the force that causes the acceleration be observable in multidimensional time? Again, we do not have enough information to know.

But what we can do is summarize these observable properties ***in 3D Time***,

1. *Observable Location in 3D Time*
2. *Observable Shape in 3D Time*
3. *Observable Motion in 3D Time*
4. *Observable Acceleration in 3D Time*

To clarify, while these listed properties are observable *in* 3D Time, none of these properties of 3D Time are observable *outside* of 3D Time. We can observe them only if we are *in* 3D Time.

From our current viewpoint in 3D Space, these four properties of 3D Time are not observable. From the 3D Space perspective, this list would be

1. *Nonobservable location in time*
2. *Nonobservable shape in time*
3. *Nonobservable motion in time*
4. *Nonobservable acceleration in time*

For the first two properties in the above list, we can visualize location and shape. Location in time is measured as a difference between the desired location and a reference location. In 3D Space, this is done in units of *distance*. In 3D Time, this is done in units of *period*.

The shape in time would be defined by volume of $seconds^3$, or maybe $minutes^3$ or $hours^3$, assuming that the period in 3D Time has the same units we have created for time in 3D Space.

Moving ***in*** time also seems somewhat intuitive, but do we really understand what that means? The same for acceleration. Let's look at these two latter properties in more detail.

6.2 Moving in 4D Time.

As we learned previously, to move from one location to another, there has to be a change in time and a change in space. Since locations and distance are defined in terms of time in 3D Time, motion has to be the change in 3D Time divided by the change in 1D Space. This is the inverse of motion in 3D Space, which is change in 3D Space divided by change in 1D Time.

velocity in 3D Time = change in 3D Time / change in 1D Space

If you, or an object, is located in 3D Time, then measuring change in location in 3D Time would also require space. Measuring *change* in 3D Time would be in units of length, for example, *meters, centimeters, or millimeters,* instead of change in 3D Space, which is in units of *hours, minutes, and seconds.* A bit of a strange concept to initially understand.

The previous equation can also be written as,

velocity in 4D Time = 1 / velocity in 4D Space

You see, there is a reciprocal relationship between the velocities in space and time. If velocity in 4D Time is 1 / velocity in 4D Space, that means,

The faster you move ***in*** *4D Space, the slower the velocity* ***in*** *4D Time. The slower you move in 4D Space, the faster you move in 4D Time.*

This relationship then conserves that amount of velocity, with the reference being the velocity of light. In the next section, particles will be introduced that have a property that, as they lose energy, they increase their speed and they can never move slower than the speed of light.

What about acceleration in 3D Time? Acceleration is the change of motion, or a change in the change in location. In 3D Space, this double change is captured in the units of acceleration, which is space / time^2, or meters / second^2. So a change in the motion in 3D Time would be time / space^2, or units of seconds / meters^2.

It takes a bit of work in your brain to begin to visualize the relationship between these two realities, but it is worth the exercise.

6.3 Speed of light in 4D Time

Another postulate of the Larson's theory is that the speed of light in 4D Time is always faster than 186,000 miles per second.

A photon moving faster than light is theorized in physics. The particle, called a tachyon, always travels faster than the speed of light. It can never go slower than the speed of light. Another particle, called the tardyon—I assume as in being tardy—can only move slower than the speed of light.

The unusual properties of tachyons are found when Special Relativity is used to describe particles traveling faster than light. Properties of these tachyons are summarized by Nick Herbert, in his book *Faster Than Light, Superluminal Loopholes in Physics:*

1. A tachyon possesses an imaginary rest mass, that is, the mass squared is a negative number.
2. As a tachyon loses energy, it speeds up. To slow it down, it has to gain energy. Once a tachyon has lost all of its energy, it must travel at an infinite velocity.
3. To slow a tachyon down to the speed of light takes infinite energy. Once a tachyon, always a tachyon.

These properties are the reciprocal of particles, labeled as tardyons, that are familiar in 4D Space. Tardyons have positive mass, so they speed up as they gain energy, and it takes infinite energy to get these particles to the speed of light.

Again, Nick Herbert states: [2]

A pleasing symmetry exists between these two types of matter–except for quantum creation events, tachyons and tardyons each stay on their own side of the speed-of-light fence.

This description is remarkably similar to the descriptions of matter in 4D Space [tardyons] and matter in 4D Time [tachyons]. So the description of matter and fields constrained to speeds below the speed of light and another description of matter and fields constrained to speeds above the speed of light is supported by theories in physics.

The only way I found to calculate this is when I evaluated the equations of electrodynamics to include 3D Time and 1D Space. These calculations I did resulted in electromagnetic waves in 4D Time that always move faster than the speed of light [Appendix A1]. The speed of electromagnetic waves in 4D Space is always less than the speed of light.

For now, given the support of theories in physics for tachyons, the faster-than-light photons, and the derivation I did in the white paper in Appendix A1, I will go forward and accept this postulate, that in 4D Time, photons always move faster than the speed of light.

6.4 Coupling Space To Time

We now know that the velocity can be used to convert time to space and space to time. We also learned that from a 4D Space point of view, location, shape, motion, and acceleration are observable since we have accepted that we live ***in*** 3D Space and ***in*** 1D Time [4D Space].

From this perspective, 4D Time location, shape, motion, and acceleration are not observable to us. They are invisible and not measurable. But could we identify these nonobservable attributes of 4D Time in 4D Space? Gravity is not observable [invisible] and not measurable without mass in 4D Space, but its effects on matter are easily measurable when we have a mass.

Let's see what we can learn by using velocity as the coupler between the dimensions of 4D Space and 4D Time to see if we can identify the nonobservable attributes of these properties of 4D Time.

The relationship between the velocity, distance, and time is as follows:

space x 1 / velocity = time

or

~~space~~ x time / ~~space~~ = time

and

time x velocity = space

or

~~time~~ x space / ~~time~~ = space

We understand what velocity and 1 / velocity can do, but what insight can velocity^2 and velocity^3, as well as 1 / velocity^2 and 1 / velocity^3 bring us?

In his Reciprocal System Theory, Dewey Larson postulates that 1 / velocity^2 has the properties of momentum and 1 / velocity^3 has properties of mass. In addition, he also postulates that 1 / velocity has the properties of energy and 1 / acceleration has the properties of force. He references the consistency of his exhaustive study of the properties of matter with regard to electricity, magnetism, heat, and energy. [3]

A few simple relationships in physics, in the form of equations, can give some insight into what these properties of velocity and acceleration in 4D Time mean, and see if the postulates regarding energy and mass can be derived.

If we want to convert motion and acceleration between space and time, we can use velocity^2, 1 / velocity^2, velocity^3 and 1 / velocity^3.

Velocity (in 4D Space) x 1 / velocity^2 = 1 / velocity (in 4D Time)

or

space / time x 1 / (space^2 / time^2) = time / space

which matches the previous outcome in section 6.2, where velocity in 4D Time equals 1 / velocity in 4D Space.

Let's look at the next equation for acceleration.

Acceleration (in 4D Space) x 1 / velocity^3 = acceleration (in 4D Time)

or

space / time^2 x 1 / (space^3 / time^3) = time / space^2

So acceleration in 4D Time has the space and time units reversed correctly. Let's rearrange the way this last equation is written.

time / space^2 *= 1 / (space*^{3} */ time*^{3}*) x space / time*^{2}

so it becomes

acceleration (in 4D Time) = 1 / velocity^3 *x acceleration (in 4D Space)*

The accelerations in 4D Time is equal to a 1 / velocity cubed term times the acceleration in 4D Space. Another important equation in physics has three terms and also uses acceleration in 4D Space, that being

Force = mass x acceleration.

After matching up the previous equation into the form of the latter using the familiar acceleration in 4D Space, we notice that force is the unobservable acceleration in time, and mass is associated with the 1 / velocity^3 term. So far, the speed of light is considered the coupler between observable 4D Space and unobservable 4D Time, so 1 / velocity^3 could be considered a coupler with properties of mass. To quote the physicist Hal Putoff: [4]

In our formulation, the m in Newton's second law of motion, F=ma, becomes nothing more than a coupling constant between acceleration and an external electromagnetic force.

Here, he considers mass to be a coupler between acceleration [in 4D Space] and force, which we now associate with an acceleration in 4D Time.

Mass, in this case, is represented as inverse of motion, or the resistance to motion, which is called inertia. The properties of 1 / $velocity^3$ fit the properties of mass, but only if velocity is the speed of light, since mass is a scalar, and if velocity is a multidimensional variable, it will make mass a multidimensional variable, which, in 4D Space, it is not. So,

$$\text{mass} = 1 / c^3$$

Let's look at another equation:

$$\text{Energy} = \text{mass} \times c^2.$$

If the equation $E = mc^2$ is solved for mass, then

$$\textit{mass} = \textit{Energy} \times 1 / c^2$$

We just correlated mass to $1 / c^3$. If we enter that into the above equation, then Energy has to be 1 / c, which is what Larson postulated. So far, it is consistent. Clearly, it holds also for the equation

$$\textit{Energy} = \tfrac{1}{2}\, \textit{mass} \times \textit{velocity}^2$$

since energy equaling mass times the speed of light squared is a special form of this equation.

Energy in 4D Space is a one-dimensional scalar variable, so it has to be 1 / speed [1 / c]. But in 4D Time, energy is equal to 1 / velocity, which in 4D Time is 3D Time divided by 1D Space. So in 4D Time, energy is a vector and has directional components.

We have found that acceleration in 4D Time has the properties of force, $1 / c^3$ has the properties of mass, $1 / c$ has properties of single-dimensional energy in 4D Space, and 1 / velocity has the properties of multidimensional energy in 4D Time. It is odd to think of time in terms of energy and force, but these equations show that the properties of space and time are interleaved and again coupled to each other by the speed of light.

The one coupler that is left out is $1 / \text{velocity}^2$, or $1 / c^2$. Let's evaluate that in the next section.

6.5 Momentum

I put momentum in a separate section, since I found that some problems occurred when equating momentum to $1 / \text{velocity}^2$. This section gets into more details about the properties of scalars, one-dimensional variables and vectors, multidimensional variables. But it is a section that is important to address mathematically. For those not so inclined, momentum is equal to $1 / \text{velocity}^2$ only when velocity is equal to the speed of light, that is, momentum equals $1 / c^2$.

For those who are interested in some of the physics and mathematical details, let's go through this exercise, and it will become clear what my concerns are.

Momentum is calculated from the variables of mass and velocity,

momentum = mass x velocity

Since mass is a one-dimensional scalar and velocity a vector, momentum is a vector. That we know from physics. Using what we have learned about mass, let's see how momentum is represented.

momentum = 1 / speed³ x velocity

This is fine, since a scalar [mass] times a vector [velocity] equals a vector [momentum]. But velocity and speed cannot cancel out, so momentum is not equal to 1/ speed²

But let's assume that the velocity is at the speed of light, so

$$momentum = 1 / c^3 \times c$$
$$= 1 / c^2$$

But there is a general dimensionality problem since momentum is a vector, so has multidimensional properties, but the term $1 / c^2$ can only have one dimension. The only way around it is if velocity is equal to the speed of light. *This means that only at the speed of light can momentum be represented as $1 / c^2$, and at this point, momentum does not have directional properties,*

To verify this, let's consider a photon. As a wave, it has energy which is equal to

Energy = Plank's constant x frequency.

We can also use this equation to define Plank's constant as

Plank's constant = energy / frequency

Energy is time / space and frequency is 1 / time, so Plank's constant is equal to time² / space.

Plank's constant = (time / space) / (1 / time)

$$= time^2 / space$$

Now the photon, treated as a particle, also has momentum equal to

momentum = Plank's constant / wavelength

Putting this spacetime value of Plank's constant into the equation for momentum, we find

$$momentum = (time^2 / space) / space = time^2 / space^2$$

which is what Larson postulated. Since photons move at the speed of light, momentum can be written as $1 / c^2$. But this cannot be generalized to momentum is equal to $1 / velocity^2$.

For now, I was not able to figure out how Larson equated momentum to $1 / velocity^2$ without putting specific limits on it. But that does not take away from the fact that $1 / velocity^2$ is an important coupling term between 4D Space and 4D Time.

6.6 Photon Angular Momentum, Energy, and Time

In Tom Bearden's book, *Energy from the Vacuum*, he discusses how the properties of time, energy, and angular momentum of the photon come together.

He states that Quantum Field Theory recognizes four polarizations [axes of motion] of photons, three in space and one in time. Angular momentum is derived from the spin of the photon and is summarized in the equation for the photon below. [5]

> *Angular momentum of photon = change in Energy x change in time*

He rephrases this equation in another way:

> *Angular momentum of photon = piece of energy in Space x piece energy in time*

Since the angular momentum of the photon is fixed at a spin of 1 [hbar], the two terms on the right-hand side are inversely related. If one goes up, the other must come down. As Bearden states,

> *...the energy and time trapped in a photon are canonical. The greater the piece of energy, the smaller the piece of time. And vice versa.*
>
> *So if one wishes to stress the 'rate of flow of time' significantly, one needs to produce large amounts of photons that have very large pieces of time and, consequently, little pieces of energy.*

The energy in 4D Space of a particle goes up with frequency. So frequencies of visible light have much more energy than sound frequencies, and gamma frequencies have much more energy than visible light frequencies.

From the relation in the above equation, it is obvious that if the piece of energy in 4D Space goes up, then the energy in the piece of Time must come down.

In order to increase the energy in the piece of Time, the piece of energy in space must be lower. The piece of energy in space is lowered by using lower frequencies. If frequency is lower, the piece of energy in time is larger.

Since the time energy increases as the frequency is lowered, whereas the space energy decreases, the total amount of energy in space and time are conserved.

If you want to affect 4D Space part of matter, use larger pieces of space energy, that is higher frequencies. If you want to change the 4D Time part of matter, use larger pieces of time energy, that is, use lower frequencies.

This insight is significant and radically changes how we think about energy and frequencies.

Bearden states that the energy in time is compressed by c^2, so the density of energy is much higher in time than in space. The basis for the much greater energy density in time that Bearden refers to comes from the equation Energy equals mass times the speed of light squared. In Bearden's book, the mass we observe in 3D Space is mass that has absorbed time, and so the nonobservable mass becomes masstime, which he states is observable. More on that topic in the next chapter.

This idea of the reciprocal nature of space and time energy is an expanded concept to everything we have been taught about energy and frequencies. If energy in time is $186{,}000^2$, or 372,000 times greater, the choice would be always to use low frequencies.

But it is clear that energy in time does not normally manifest directly into energy in space. Under the right conditions, it can be made to happen. One of those conditions is atomic energy, where Bearden states that energy from time flows into 4D Space, but at the increased density of c^2. So if you have a choice, create conditions that use energy from the time domain.

Given this new insight, we learn that sound frequencies have much more energy in time than do visible light frequencies. What does that say about our vocal frequencies? We know that the spoken word is very powerful. It has the power to move whole masses of people, but as we currently understand it, it does not have the power to move mountains. Or could it?

6.7 Frequency and Structure

In considering 4D Time, one of the properties associated with time, is frequency, or 1 / time. But thinking of 4D Time as having structure can seem a bit strange at first, so, in this section, we will explore how the property of frequency is associated with shape and structure.

Mass is made up of energy, and energy is based on frequencies, so all the matter we know is based on frequency.

If we take two common physics equations used previously,

$$Energy = mass \times c^2$$

and

$$Energy = Plank's\ constant \times frequency$$

And equate the two and solve for mass

$$Mass = (Plank's\ constant\ /\ c^2) \times frequency$$

Since Plank's constant and the speed of light are constant, all the mass we know of is based on frequency.

Music is a time phenomena. The beat is a period measured directly in time, and frequency is measured in inverse time, that is 1 / time, or 1 / seconds, or Hertz.

Music is also ratios of time. Chords are combinations, that is ratios of one frequency relative to others. These ratios have specific effects on the structure of the music as well as the listener. Mathematics, via the ratios of frequencies, is very important to the structuring of a musical composition.

Frequency, when interacting with matter in Space, produces specific vibrational patterns in matter. These patterns, due to matter resonating with specific frequencies, can be seen as geometric patterns, known as Chladni and Cymatic patterns,which are shown in images below and on the next page. The Chladni patterns clearly show how frequency can influence shape. The interface is the mass, in the case of Chladni, a metal plate, and in the case of Cymatics, a fluid. The Chladni patterns change dramatically when the frequency is changed by just a small percentage.

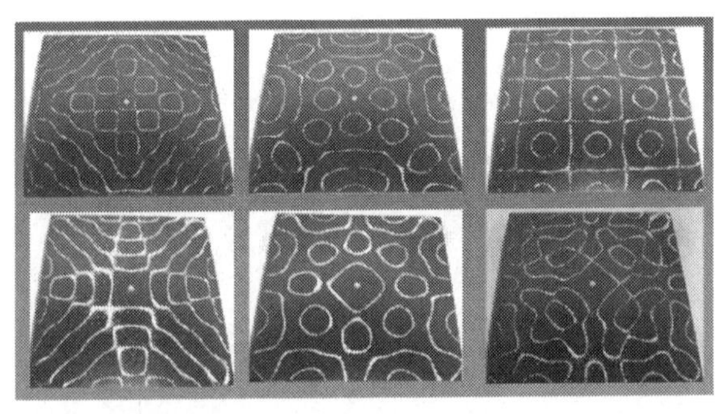

Chladni Patterns [Author's photos]

In Cymatics, when fluids of different composition and densities are excited resonantly with frequencies, the geometric patterns are amazingly complex, as the Cymatic images show. Some of these patterns can be made to look like biological life structures, as shown in the third Cymatic image below [6]. In movies of these structures, the patterns move as if they are alive.

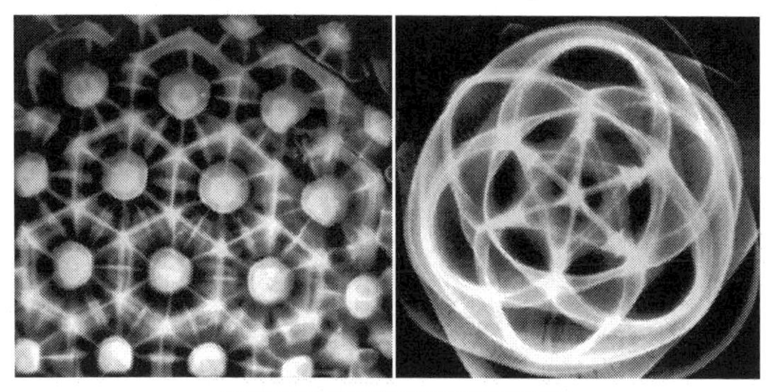

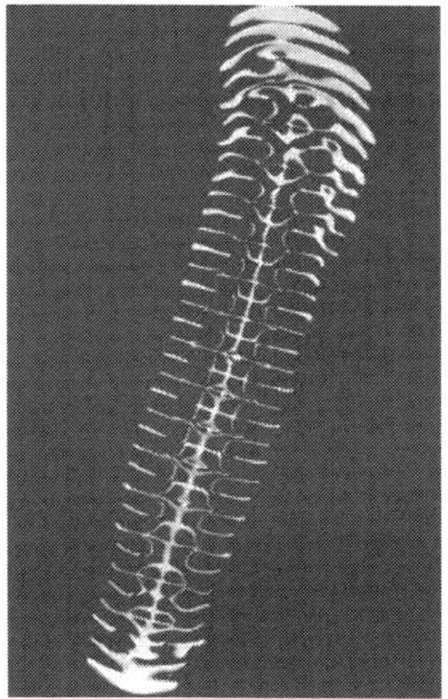

Reference Cymatics [6]

6.8 1D Space in 4D Time

What would the space dimension of 4D Time look like? It should have the same nonlocal wavelike properties as 1D Time has in 4D Space. The way I think of it is as a very uniform distribution of matter, like the density of air. To visualize this, think of a fog that is everywhere in three-dimensional time.

Another way to visualize this is the distribution of air molecules, or the density of air, is coupled at a particular temperature. This temperature can also be thought of as energy. The atmosphere is filled with a distribution of air molecules that have a specific uniform density at each temperature. During the day, the temperature heats up, the density decreases, and at night, it cools down and the density increases. But the uniformity of the air for a *given temperature* stays relatively the same. So the density of 1D Space can go up and down, just like 1D Time in 3D Space.

6.9 Properties of 4D Time.

1. The first property of 4D Time is shape and location. Correlates to periods in time. Units of T, T^2, and T^3.
2. All motion in 4D Time is faster than the speed of light. The faster the motion in time, the slower the motion in space. The two motions are bound by this ratio known as the speed of light.
3. The second property of 4D Time is energy. Energy in 4D Time correlates to inverse velocity. Units T / S or Time / Space. Energy in 4D Time is a vector.
4. The energy of photon in space and time are reciprocal. High-frequency photons have large energy in 4D Space, but low energy in 4D Time. Low frequencies have large energy in 4D Time and low energy in 4D Space. Energy is conserved when energies in both 4D Space and 4D Time are considered.
5. The third property of 4D Time is Force. Correlates to reciprocal acceleration. Units T / S^2 or Time / $Space^2$.This is associated with the coupling 1 / $velocity^3$ [mass].
6. 3D Time is the local part of these extra dimensions of spacetime and so is associated with the particle-like nature of matter in 4D Time.

7. 1D Space in 4D Time is everywhere in the same manner that 1D time is everywhere in 4D Space. So 1D Space is nonlocal, that is, it is the wavelike nature of matter in 4D Time.

7 Four-Dimensions of Space and Four-Dimensions of Time.

Time has been transformed, and we have changed; it has advanced and set us in motion; it has unveiled its face, inspiring us with bewilderment and exhilaration.

Khalil Gibran

So far, we have discussed the familiar 3D Space and 1D Time, and hopefully, the other dimensions of 3D Time and 1D Space are a bit more understandable now.

Now, both space and time are not tangible. We cannot reach out and touch either. It does seem odd to say we cannot touch space, or measure space, never mind time. But what we can do is reach out and touch *matter*, which is mass located in both space and time, and as we saw in the previous section, mass is defined as the 3D coupler between space [acceleration] and time [force]. Remember, mass is motion cubed, so without motion, there is no space or time, and certainly no matter.

But our construction of 4D Space allows for easy identification of this matter in Space. We can come back time and time again to a huge boulder resting on the surface of the earth, and it will still be there. We cannot do this for time. We can identify the rock in each moment of time, then once that moment in time has passed, we have to rely on memory of the past, often in the form of recorded images, and imagination to project its fate into the future.

We associate the particle and local nature of matter to 3D Space, and the wave and nonlocal entangled nature to 1D Time.

We then mapped out another dimensional reality, 3D Time and 1D Space. In this reality, locations in time can be traveled to again and again, and an object in 4D Time will be at that location. As discussed, just like time in 4D Space, in 4D Time, it is space that is everywhere uniform. 1D Space is nonlocal and entangled in 4D Time. Odd to think in those terms, since we are so familiar with time in 4D Space.

So 1D Time in 4D Space and 1D Space in 4D Time are the nonlocal, or the wavelike nature of matter. This means that matter, or motion, is everywhere in the universe and connected through 1D Space and 1D Time.

Velocity can have extra dimensions of space or time, but not speed. 3D Space and 3D Time are the local, or particle-like, natures of matter or motion, but are not coupled in 3D, but only in 1D. So 1D Time has to be expanded into 3D Time, and 3D Space has to be compressed into 1D Space to couple all these dimensions together. Luckily, there is a way to do this, which will be explained in the next section.

In 4D Time, as discussed in the previous chapter, the objects have shapes and locations determined by time, so what we associate shape with will be completely different. If you change the frequency, the amount of time changes. Change the energy content and the shape changes. But like time in 4D Space, the shape in 3D Time can only be defined at one moment in space. Shape in time can only change if space changes along with a change in time, which requires energy. Energy is inverse velocity, which is a change in time divided by a change in space.

7.1 Transforming 3D to 1D and 1D to 3D

In these two domains, we have two components of space and two components of time. In 4D Space we have 1D Time, and in 4D Time we have 1D Space.

In the Reciprocal System Theory the explanation of the coupling between 3D Space and 1D Space, or 3D Time and 1D Time, is the speed of light.

How do 3D Time and 1D Time, as well as 3D Space and 1D Space, relate to each other? We have shown clearly that they have very different properties, even though they are the same variable.

We know that 3D Space and 3D Time have vector properties, and so have directional properties, whereas 1D Space and 1D Time do not. In mathematics, there is an operator called 'divergence,' which compresses three-dimensional variable into a one-dimensional variable. So 3D Space can be converted into 1D Space, and 3D Time can be converted into 1D Time. But what about the other way around?

There is another operator, called 'gradient' that takes a one-dimensional variable and turns it into a three-dimensional variable. 1D Time can be turned back into 3D Time, and 1D Space can be turned into 3D Space.

In this way, there is the ability to have feedback between 1D and 3D Space and 1D and 3D Time. More details of this feedback between 1D and 3D is captured in Appendix A1, where the discussion of these 1D and 3D variables in relation to electrodynamics are discussed.

7.2 How does it all fit together?

As we mentioned before, one of the mysteries is how the electron, spinning around the atomic nucleus for billions of years at the speed of light, never runs out of energy. The physicist Hal Putoff theorized just the right amount of energy is transferred over billions of years to each and every electron in the universe, an incredible amount of electrons, preventing these electrons from a death spiral into the nucleus due to the electromagnetic pull of the nucleus. Now the key here is 'just the right amount of energy.' If too little energy transference happens, the electron cannot overcome the electromagnetic pull of the nucleus, and electrons would disappear into the nucleus. If there is too much energy transference, the electrons will slowly escape from the same electromagnetic pull of the nucleus. In each case, the properties of atoms and molecules will be forever changed. [1]

Dewey Larson, in his theory of matter, has stated that 4D Space and 4D Time [he denotes these as timespace and spacetime] do not exist at the same time, but that motion is an oscillation between these two-dimensional realities. But Larson is not the only one. Tom Bearden also describes reality as oscillating between two states, but in his model, it is between mass and masstime, a topic discussed in the next section.

Remember, in theory, this oscillation could be at 10^{43} Hertz. So what looks like simultaneous events in multidimensional 4D Space and 4D Time can easily be supported by this extremely high oscillation. Just like a movie, the reality viewed is seen as continuous, since we cannot detect the oscillation between static images. Since we cannot even remotely come close to measuring this oscillation of multidimensional 4D Space and 4D Time, in all our scientific investigations, multidimensional 4D Space and 4D Time appear instantaneous. So if we are oscillating between two realities at such a fast rate that both appear simultaneous, it means we are experiencing an eight-dimensional spacetime.

But let's look at these transitions between 4D Space and 4D Time universes closely, so that we might understand the relationship between them. That means that at any point, we are either in the 4D Space universe [observable 3D Space and unobservable 1D Time] or the 4D Time universe [unobservable 3D Time and unobservable 1D Space]. The fact that so much of this eight-dimensional spacetime is unobservable might be used to explain why so much of matter and energy is dark. More on that later.

In the first state, the observable local particle-like 3D Space and the unobservable nonlocal wavelike 1D Time associated with 4D Space universe is active. Here, 1D Time generates change in 3D Space through motion, that is 3D Space divided by 1D Time. This motion in 3D Space is the only part that is observable to us [black arrow pointing up shows direction of change].

In the next state, the unobservable local particle-like 3D Time and the unobservable entangled nonlocal wavelike 1D Space associated with 4D Time is active.

How does this first transition happen between 4D Space and 4D Time universes? In this first transition, two actions happen simultaneously. In one action, all the information of motion from 3D Space is captured and compressed into 1D Space. This information of space is now nonlocal, and so is entangled and available everywhere in the unobservable 4D Time universe. From our perspective in the 4D Space universe, both are unobservable. In the second action, the nonlocal 1D Time information from 4D Space is expanded into 3D Time information in 4D Time universe.

To visualize these transitions, it might be easier to also follow along the progression shown in Table 1 below. So in this second transition, space is compressed from three dimensions to one dimension [black arrows pointing bottom right], and time is expanded from one dimension into three dimensions [gray arrows pointing top right]. Here, motion, or change, is created by 3D Time interfacing with 1D Space in the form of motion. This motion is 3D Time divided by 1D Space, or relative to us in 4D Space, it is 1 / velocity in space.

Universe	4D Space	4D Time	4D Space	4D Time	4D Space
State	Observable	Unobservable	Observable	Unobservable	Observable
Motion	3D Space / 1D Time	3D Time / 1D Space	3D Space / 1D Time	3D Time / 1D Space	3D Space / 1D Time
How change from 1D manifests in 3D.	Here 1D Space from previous step [not shown] expands into 3D Space	Here 1D Time from previous step expands into 3D Time [gray arrow].	Here 1D Space from previous step expands into 3D Space [gray arrow].	Here 1D Time from previous step expands into 3D Time [gray arrow].	Here 1D Space from previous step expands into 3D Space [gray arrow].
Local	3D Space	3D Time	3D Space	3D Time	3D Space
Arrows of change					
Nonlocal	1D Time	1D Space	1D Time	1D Space	1D Time
Agent of change	In this step, 1D Time changes 3D Space. [black arrow]	In this step, 1D Space changes 3D Time. [black arrow].	In this step, 1D Time changes 3D Space. [black arrow].	In this step, 1D Space changes 3D Time. [black arrow].	In this step, 1D Time changes 3D Time. [black arrow].
How information in 3D is available everywhere in 1D	In next step, information from 3D Space compresses into 1D Space.	Information from 3D Space is compressed into 1D Space – it is now available everywhere in 3D Time.	Information from 3D Time is compressed into 1D Time – it is now available everywhere in 3D Space.	Information from 3D Space is compressed into 1D Space – it is now available everywhere in 3D Time.	Information from 3D Space is compressed into 1D Space – it is now available everywhere in 3D Space.

Table 1. How it all fits together

Remember, the slower you move in 4D Space, the faster you move in 4D Time.

Now, the second transition is out of 4D Time back into 4D Space. The information, which has changed during its time in unobservable 4D Time, is transferred into 4D Space. In 4D Time, change acted on 3D Time by interfacing with 1D Space through motion. When the changed information in 3D Time is compressed into 1D Time, these changes are now nonlocal and available everywhere in the 4D Space universe. The unobservable entangled information from 1D Space is expanded back into observable 3D Space. Now the agent of change is motion, which is 3D Space divided by 1D Time, but this 1D Time has been modified by entanglement while in the 3D Time state of the 4D Time universe.

To understand the flow of change, let's follow how change happens. In 3D Space or 3D Time, change happens with motion in a localized way. In 1D Time and 1D Space, change happens via entanglement, which is distributed motion. In the 4D Space universe, 1D Time is the agent of change for 3D Space through local motion. In the transition to the 4D Time universe, this change in information is captured and compressed into 1D Space, and through entanglement, it interfaces with all other motion. Think of a three-dimensional matrix of balls, where each one is connected to surrounding balls. In 1D Space, through entanglement, every part of the universe interacts with every other part of the universe. This 1D Space information is the agent of change for 3D Time via motion, but, in this case, time divided by space.

The second transition back to the 4D Space universe again, the nonlocal entanglement 1D Space from the 4D Time universe is expanded into 3D Space, but with changes from the entangled 1D Space state. The changes made on 3D Time by 1D Space while in the 4D Time universe is compressed into the entangled nonlocal 1D Time state of the 4D Space universe, our familiar version of time. So when 3D Space is divided by this potentially changed 1D Time, there is the potential for a new motion in 4D Space.This switching is happening at extremely fast frequency.

One symbol that I feel captures this integration of 4D Space and 4D Time very well symbolically is the yin yang symbol.

If we have the white part represent space and the black part represent time, this symbol represents the following to me: The black dot in the white space represents 1D Time in 3D Space. The white dot in the black space represents 1D Space in 3D Time. So the 3D black time is coupled into the 3D white space via the black dot representing nonlocal time. The reverse would apply for 3D black time and the 1D white dot. The two domains of 4D Space and 4D Time are swirling around because motion is the interface between these two. More on this motion of spin in the next chapter.

Everything we are made of, our bodies and everything in the world around us, is oscillating between these two universes. If the building blocks are oscillating, so are we as humans. Remember the analogy of our eyes, without motion we cannot see, but we are not aware that our eyes are constantly moving at around three Hertz all day, every day. So it seems our sense of the universe appears continuous, but it is oscillating at some extremely rapid rate between 4D Space and 4D Time.

I am not sure if the duty cycle would be exactly 50%, that is we move 50% in one reality and in the other. Notice I did not say half the time in one and half the time in the other, since time is a property of both of these realities, and the fundamental reality underlying both of these universes is ***motion***. Since we primarily identify as beings moving through 4D Space, a movement of 50% in multidimensional Space is equivalent to a movement in multidimensional time of 1 / Space. The same movement, but the experience of the movement in time would be very short.

Dewey Larson really stresses motion to help get us out of our common perception of 4D Space. With motion, space and time are automatically generated. What generates the original motion is another topic. I believe that it is Zero-Point Energy that generates this motion. The Zero-Point Energy is derived from the Zero-Point Field, but what created the Zero-Point Field? For now, there are enough brain gymnastics just to get these multidimensional realities clear, so we will stop at the Zero-Point Field as being the primal cause of motion.

Naively, the default assumption is that the setting for duty cycle between 4D Time and 4D Space would be 50%. But should it be limited to that? What if you could change that? As you increase the time duty cycle, I would expect that more and more matter should become invisible to us, since time is not observable to us. Also, does matter become less dense, more gaseous or liquid-like as time increases? There is no way to be sure. Nature is always moving to a place of balance, so it is likely that the balance of 4D Time to 4D Space is manifested in a way that allows just the right amount of energy to come out of the Zero-Point Field to support our universe as we know it.

If Bearden's statement that energy density in time is 372,000 times greater, maybe that might lead to some insight as to why in the ratio we call the speed of light, we have such a large distance [space] for a relatively small increment period [time]. The speed of light in our material universe might have to be this way to offset this huge energy density in time. The theory of Stochastic Electrodynamics shows that even a small change in the balance of the universe would result in huge changes, given the incredible amount of potential energy locked up in the Zero-Point Field. So, thankfully, these variables are remarkably balanced and constant over time.

7.3 Mass and Masstime

In this section we go back to one of the couplers, mass, between force and acceleration. In a previous section, mass is correlated to $1 / c^3$. So mass is inverse motion in three dimensions. But what is this phenomenon that we observe called mass? To explore this concept in more depth, I found it useful to evaluate a model of mass that Tom Bearden has put together.

In his model, mass is independent of time. When a mass, which we cannot see, absorbs a photon it becomes masstime. Now the photon is a piece of energy in space and time. This state of masstime is what science recognizes as observable mass. Before the mass absorbed the photon, with its piece of time, it was unobservable mass.

When the photon is emitted from this masstime state, this photon carries with it a piece of time changed by the properties of this mass. Again, mass is just motion in three dimensions. This photon is what we are measuring, and therefore, making the mass observable to us. He states that without this photon, the mass is not observable to us. This constant oscillation between the discrete states of mass [m] and masstime [mt] is happening at an extremely fast rate, but we experience this as a persistent flow of time, in the forward direction. His example is one of a movie, where the movie is just a series of discrete static images, but we experience it as a continuous reality. We can only experience the movie when a static image is illuminated with photons of light. So it is with mass—we only see, or measure, the photons being ejected from the masstime frames, making the mass observable to us. [2]

Bearden states,

A major point is that mass does not emit a photon; masstime does. Mass 'travels through time' by an extremely high oscillation between corpuscle-like state m and wavelike state mt…

This oscillation between corpuscle-like [particle] m states and wavelike *mt* states is so fast that both states are available to us at the same time. Now, Bearden always talks about time as 1D time, so I will put what he has said here in perspective of 3D Time.

Remember, time and space are not tangible or observable to us. Since motion is the foundation of time and space, this motion is unobservable. Previously, mass or matter is correlated to motion in three dimensions, but this motion, as stated above, is not observable, making the matter unobservable.

In the context of 3D Time, I visualize that when mass absorbs an 'amount of energy from time' in the form of a photon, it is from 3D Time, so mass becomes masstime. Then the masstime in 3D Space emits a photon that has a piece of energy in space and energy in time, but this time is the 1D time we are familiar with. It is during this moment that masstime has ejected a photon in 4D Space that the mass is observable to us and our instruments. Remember, mass is equivalent to frequency times a constant. This constant is Plank's constant divided by speed of light squared. As the masstime releases this time as a unit of energy in the form of a photon, it goes back to being mass again.

Immediately, the mass absorbs another 'piece of time.' So it is the act of absorbing time that generates the change, and releasing this piece of time [photon] in 3D Space is the information we use to detect this change of the mass in time.

7.4 Charges – source of infinite energy

Tom Bearden is fond of saying that creating a dipole, a positive and negative charge separated by some distance, is breaking the symmetry of the vacuum. When this is done, time streams into 3D Space. His model is one where the time domain, which he defined as the familiar 1D time, is embedded in the Quantum Vacuum, or Zero-Point Field. Tom Bearden does not talk anywhere about multidimensional time, although his insightful ideas are not limited by the idea of multidimensional time.

This energy from the time domain flows as long as the dipole exists, possibly indefinitely [3]. So the electric fields associated with the positive and negative electric charges outflow at the speed of light. In two seconds, these fields encompass a sphere much larger than the distance of the earth to the moon. The electrons have been outpouring this energy for billions of years. In fact, if it were not for these electric fields, all our structures of matter would collapse. Our matter, which is more than 99% nothing, is bound together by the attractive and repulsive forces between these electromagnetic fields.

Bearden states that the observable energy flow is not conserved in 3D Space, since the conservation of energy includes time. This statement is significant.

Bearden goes on to say [4],

The energy steadily pouring out of the dipole in all directions in 3 Space must therefore be received by the dipole charges from the time domain.

In physics, this is the same as receiving energy from the Zero-Point Field. In a later chapter, I will show how 4D Time has many of the properties associated with the Zero-Point Field.

7.5 Energy and charge in 4D Time and 4D Space

If energy must be conserved using both space and time, and energy flows from the 4D Time domain for a negative charge, then there must be a mechanism where time can flow back into the space domain. The best material I have read on this topic is again Tom Bearden. Quoting his book [5]:

We now point out something deeper and very important. In modern physics, the observed positive charge is regarded as the time reversal of the observed negative charge. If the negative charge is responsible for EM energy flow from the time-domain into 3-space, then as observed the positive charge must represent a time-reversed situation . . .

When observed, both the sign of the positive charge and its direction of motion appear to us to be reversed from the negative case. It follows that the positive charge as observed involves a flow of (positive) energy from 3-space back to the time domain . . .

Thus, a dipolarity involves a continuous flow of energy from the time domain into 3-space, and back from 3-space into the time domain.

Note that Bearden refers to the positive charge in the time domain as a time-reversed negative charge in space. The importance of referring to the positive charge as time-reversed will be discussed later in this section.

The positive charges in 4D Space [he refers it to as 3-space] are *sinks* of electromagnetic energy from 4D Time to 4D Space. Negative sources are *sinks* of electromagnetic energy from 4D Space back into 4D Time. A *source* is one where energy is outpouring, and a *sink* is one where energy is pouring in.

From the perspective of 4D Time, negative charges are sinks of energy from 4D Time into 4D Space, and positive charges are sources of electromagnetic energy from 4D Space into 4D Time.

To continue with his analogy [6]:

The negative charge absorbs a little bit of positive time energy, transduces it into a much larger amount of 3-space excitation energy, and re-emits it in 3-space as an observable, real flow of EM energy radially outward in all directions. The positive charge receives the large amount of 3-spatial EM energy and absorbs it, transduces it back into time-energy (highly compressed energy) and re-emits it in the time domain as a little bit of highly compressed time energy.

Conversely, we may consider the positive charge absorbs a little bit of negative time energy, transduces it into a much larger amount of negative 3-space excitation energy, and re-emits this negative 3-space energy in all directions in 3 space.

With this new understanding of the conservation of energy, energy can flow back and forth between 4D Space and 4D Time so that it remains conserved. In addition, the conservation of charge would include 4D

Space and 4D Time, so that for every observable charge in 4D Space, there must be a nonobservable charge in 4D Time. It also means that if the speed of the charge in 4D Space increases, the linked unobservable charge in 4D Time has to slow down.

7.6 Electrodynamics in 8D Spacetime.

With the new understanding of conservation of charge and energy, what would change if electrodynamics is considered in eight dimensions of spacetime, 4D Space and 4D Time? This question plagued me for years, and I spent time working on what the implications would be. When I had derived a satisfactory answer, I captured it in a white paper "Concepts of Three-dimensional Vector Time in Electrodynamics." In this section, I will discuss the interesting results. [Appendix A 1].

In 4D Space, the electric fields are derived from individual positive and negative charges, called monopoles, because they exist independent of the other. Magnetic fields, on the other hand, are derived from magnetic moments, which always include a north and a south pole. No matter how small matter is divided, the north and south pole are never separated. So the magnetic field is from a flow, called a flux, between the north and south pole. Magnetic monopoles should exist, but they have never been measured.

In physics, if electric and magnetic fields are considered in a vacuum, devoid of matter, the electric field and the magnetic field have the same form of the equations to describe them. But when mass is introduced, then dissimilarities occur. Electric charges are still considered as monopoles, but there are no corresponding magnetic monopole charges, only a flux between north and south poles.

When 4D Time is introduced and the equations for electrodynamics for 4D Time are derived, an interesting result occurs. In 4D Time, the magnetic field has the units of charges, and electric field has the units of flux. If this is correct, this means that in 4D Time, no matter how small you divide matter in 4D Time, the positive and negative charges can never be separated, just like the magnetic north and south pole in 4D Space. In addition, in 4D Time, the magnetic charges are independent, so you have magnetic monopoles in 4D Time. So in 4D Time, the electric field and magnetic fields are complementary in behavior to those in 4D Space.

Also, when the speed at which the electromagnetic fields propagate are calculated, the electromagnetic fields in 4D Space are slower than the speed of light. For 4D Time, the speed for the electromagnetic fields is always greater than the speed of light.

To prove this theory of magnetic monopoles in 4D Time is difficult. Since the results of the calculations have the magnetic monopoles in 4D Time, they are not observable. Remember, not being observable means they cannot be measured. But maybe they leave traces of their existence, much like antimatter particles leave momentary traces of their existence.

7.7 Electric 4D Space and Magnetic 4D Time

From the previous section, the results of the derivation are that only electric monopoles exist independently in 4D Space, and only magnetic monopoles exist independently in 4D Time. We know that in 3D Space, you need an electric current to generate a magnetic field. An electric current is the movement of charges, that is, the change in charge over a period of time. So the introduction of time into

3D Space generates magnetic fields. *Without time, there cannot be magnetism*. Even magnets have their magnetism derived from either a spinning charge or a moving charge between a positive and negative electric pole. This can be summarized as,

Electric monopoles are the basis for electromagnetism in 4D Space.

When magnetism is generated, it automatically appears paired with two poles, north and south. It is always paired, no matter how small matter is divided.

In the previous section on charges, the idea was introduced that the conservation of charge had to include not only charges in 4D Space, but also in 4D Time. If charge is separated so that an electric dipole is created in 4D Space, it automatically generates an electric dipole in 4D Time, but with a negative charge in 4D Time linking up with the positive charge in 4D Space and a positive charge in 4D Time linking up with the negative charge in 4D Space. But it is always an electric dipole in 4D Time. You can *never* separate out the positive and negative electric poles in 4D Time, in the same way that you cannot separate a north pole from a south pole in a magnetic dipole in 4D Space.

In 4D Time, the north and south poles are independent, and it is the motion of magnetic charges in 4D Time that generates an electric field. *Without space, there cannot be electric fields in 4D Time*. So,

Magnetic monopoles are the basis for electromagnetism in 4D Time.

When a magnetic dipole is created in 4D Time, the north pole in 4D Time will automatically generate a south pole in 4D Space, and the same for the south pole. The conservation of electromagnetic energy is achieved over both 4D Space and 4D Time domains.

7.8 *Shape of electromagnetic fields*

In 4D Space, electric charges [electric monopoles] have radial fields when in total isolation.

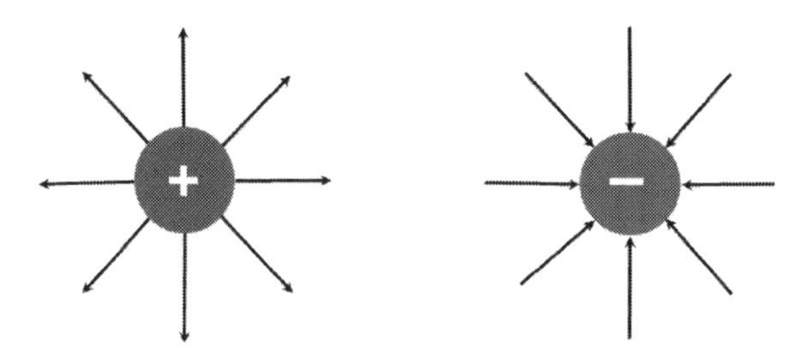

Electric fields of isolated electric charges

When a pair of positive and negative charges are in close proximity as in a dipole configuration, the field lines of this electric field are in the shape of a toroid.

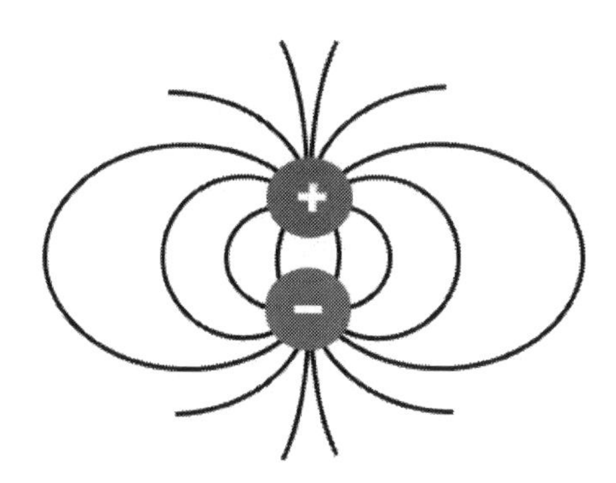

Field for an Electric dipole

For a magnetic dipole, the field has the same shape.

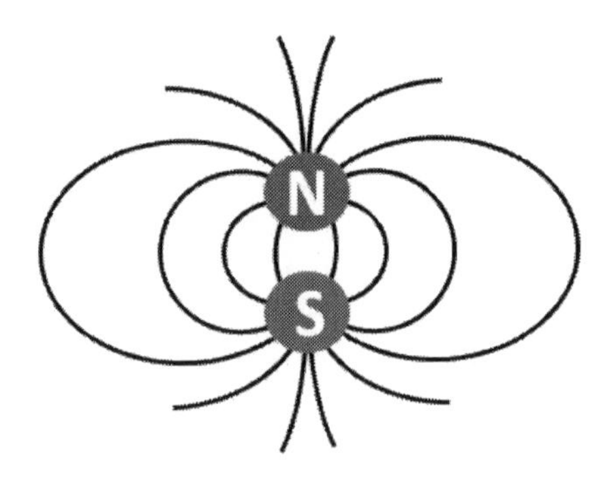

Field for a Magnetic dipole

When two poles, be it electrical or magnetic, are separated, there is a flow of energy between 4D Time and 4D Space. Let's use the electrical dipole to visualize this. The energy in the electric field flows out of the positive charge into 4D Space to a space value, or distance, that could be up to infinity and from 4D Space back into the negative charge. In 4D Time, the exact same process is happening, but now out of the 4D Time negative charge to a time value, or period, that could be up to infinity and from 4D Time back into the 4D Time positive charge.

Let's try to draw this out. We start from the beginning, with a balanced Zero-Point Field, represented as a dot, a perfect circle, on the left. Here, positive and negative charges balance each other out. Now let's consider an electron in 4D Space. It has to have a positron partner in 4D Time to balance out the total charge in the Zero-Point Field.

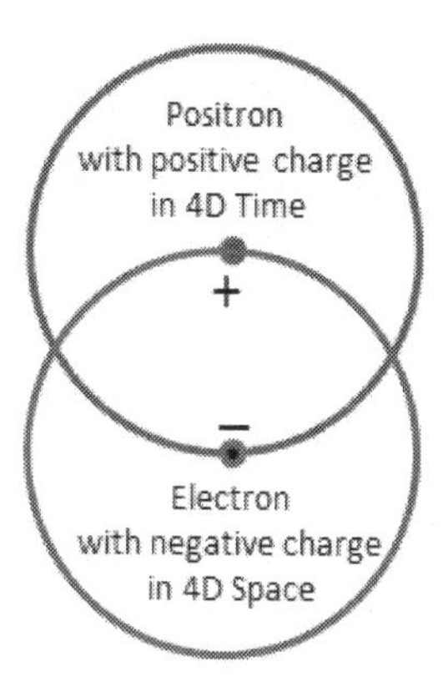

I chose to draw this dipole with the Vesica Piscis format because of some interesting symbology. The electron has a location in 3D Space, and the single dimension of the line of the 4D Time circle intersects with the electron in 3D Space, completing its dimensions in 4D Space. The same is true for the positron, which has a location in 3D Time, and has a single dimension line from 4D Space intersecting it, completing its dimensions in 4D Time. The Vesica Piscis, the overlap of 4D Space with 4D Time represent the interface, and interaction, between these two domains. This is the overlap of 1D Space with 1D Time, or the speed of light.

It should be noted that the positron is considered a time-reverse electron, so the coupling between the positron and the electron is through time. In the first chapter, the properties of time reverse was discussed, so whatever properties the electron has, the positron has the time-reversed properties so that when these two particles come back together again, they can join back into light.

But what happens when we generate a dipole in 4D Space? We have positive and negative charges, with their associated radial fields for both 4D Space and 4D Time. I visualize them interfacing as shown in the next diagram. We know the field has the shape of a toroid in 4D Space, and that the polarity of the electric poles are reverse in 4D Time, but what about the shape of the fields in 4D Time?

If the toroid shape is a 4D Space property of electric and magnetic dipole fields, then the shape of these same fields in 4D Time could be anything. If the assumption is that the shape of the fields is due to their electromagnetic nature, then the fields in 4D Time are also toroids, but with units of time instead of space.

In addition, since the fields in 4D Time are unobservable, whatever shape these fields might have, we cannot verify what it is. So for now, I use the circles as representations of charges in 4D Time. The representation of charges in 4D Time are shown as dotted circles below.

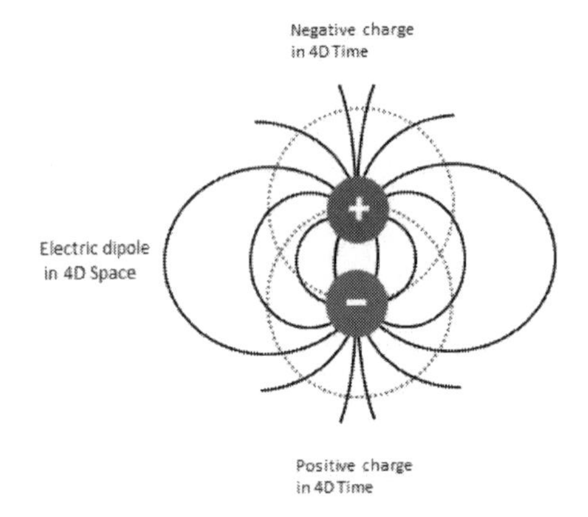

To visualize the toroid shape of the dipole in three dimensions of space, a nice representation of a torus shape is shown on the next page.

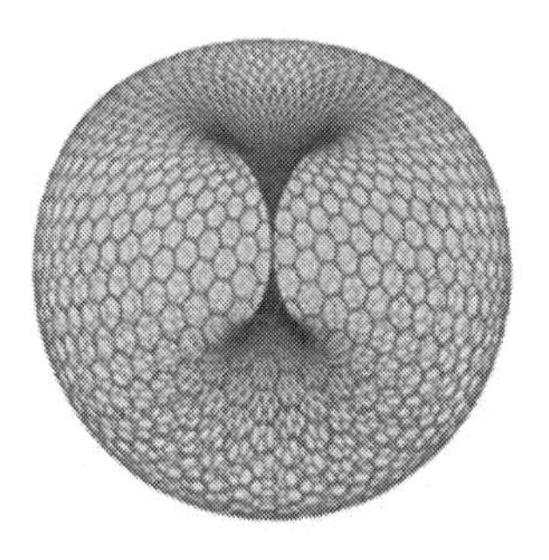

Reference. www.horntorus.com

The charges in 4D Space have spin ½, therefore they also have spin angular momentum. If you add rotation to the toroid shape, the toroid becomes more dynamic than just a static toroid. The universe is based on motion, and so there is motion in these toroids in both axes. The toroid in fluidics, like smoke rings, have rotation around the ring axis. The ring axis rotates perpendicular to the rotation around the central axis. This is the reason that toroidal smoke rings move, because of the difference in air pressure between the outer and inner surfaces of the toroidal ring.

When you add rotation around the center axis, the toroidal system becomes even more dynamic, and the surface flows in a spiraling motion from outer surface to inner surface of the toroid.

There are many Websites on the Internet that have very creative models that change dynamically in time in all sorts of ways, and I encourage the reader to do their own investigations into the possible shapes of eight-dimensional spacetime electromagnetic fields.

7.9 Dark Matters

The universe as we know it, so far contains only five percent [4.9%, actually] visible matter. Of the rest of the matter, 27% is called dark matter and 68% is called dark energy.

Dark matter is postulated by astrophysicists to account for the large amount of mass that is required to justify the gravity necessary to explain the velocities of rotation and gravitational lensing of the large objects in our universe. The amount of visible matter could not account for this, so the hypothesis is that there has to be invisible, or dark matter, that we are unable to detect or see.

A CERN webpage spoke of a theory of dark matter being extra dimensions, especially dimensions that are smaller than an atom. This seems to fit well with 4D Time, since dimensions of space that are very small, like in an atom, are places where the motion is in 4D Time.

Dark Energy, different from Dark Matter, is also postulated by astrophysicists to account for the fact that the universe is expanding. But not just expanding, but expanding at an accelerating rate. To account for this expansion of the universe, there has to be a source of energy that can sustain this expansion.

With the new view of eight-dimensional spacetime, as space expands, time contracts, and as space contracts, time expands.

If 3D Space expands, as astrophysicists are now measuring in our observable 4D Space universe, then in 4D Time, the opposition will be a contracting force, a force we could call gravity.

With the 4D Space and 4D Time model, this balance of expansion and contraction is seen to be conserved between these eight dimensions. If the expansion in 4D Space slows down and reverses, as is the expectation of astrophysicists, then in 4D Time, the contraction has to slow down and reverse into an expansion in order to conserve this balance.

This Dark Energy could just be the contraction reaction in 4D Time to this expansion in 4D Space. In the eight-dimensional model of spacetime, this expansion is the result of the conservation of expansion/contraction, a balance toward unity.

The question then becomes, what is the driver that is causing 4D Space to expand and 4D Time to contract? Is it part of a greater cycle like so many cycles in nature?

As was discussed earlier, the energy density in time is the speed of light squared higher than that of space. With 372,000 times more energy, if the amount of matter was equal parts of space and time, then 4D Time would account for so much more than 95%. It would appear that only a portion of the energy density of time manifests into 4D Space matter. This limitation could be like the limit on the number of frequencies that the Zero-Point Energy can support, a limit that is necessary to keep the universe in balance. This limit is discussed in more detail in the next section.

In addition, with multidimensional time associated with magnetic charges or monopoles, and now gravity, it could mean that time, gravity, and the source magnetic charges could all be correlated in some yet-to-be-discovered way.

7.10 4D Time, Dark Energy, and Zero-Point Field

Physicists have proposed that while the frequency limit for the Zero-Point Field [ZPF] can be as high as Plank's frequency, there is likely a cutoff frequency. If the energy density of visible, and dark energy is taken into account, and an upper limit of frequency is calculated, it is in the low 10^{12} Hertz range [7].

Beck et al have proposed that with a cutoff frequency in this range, the energy density of Zero-Point Energy [ZPE] matches that of the energy density of dark energy, so it could be that the ZPE, which has many of the same properties of time is the source of this dark energy as well. [8]

It does not seem unreasonable to use this same idea of a cutoff frequency in relation to the energy density of 4D Time, since frequency is a property of time. This cutoff frequency would act as a limit of how much of the 3D Time Field energy density manifests in the 1D Time Field. This limit acts as a gate valve for energy to flow out of time.

The extreme uniformity of the time, and of ZPF, means that there is a balance being maintained in the universe. Without the balance, the gradients would manifest a force that would rip matter as we know it apart. The model of 4D Space and 4D Time allows this model with forces of contraction and expansion and the *balance* of these two forces *moderates* the expansion or contraction of matter.

There are many properties of multidimensional time that correlate well with the properties of the Zero-Point Field. Four of the correlations are listed below.

1. Motion: The atom is always in motion, always, due to the relationship of space to time. In the theory of the Zero-Point Energy, this motion of the atom manifests from the Zero-Point Field. This motion due to ZPE of the atomic parts, at the most fundamental level of matter, could be the generator of spacetime as we know it, since motion is the basis of spacetime.
2. Shielding: Time pervades all spacetime, and it can't be shielded against, so it shares this property with Zero-Point Field (and gravity too).
3. Small distances: Since space and time are reciprocal, when space is very small, like atomic

dimensions, the time component is very large. ZPE is much more intense at very small distances, due to cubed frequency dependency. Time, again, is associated with energy, so ZPE and time behavior in 4D space are similar in that the most energy is manifested in small spatial dimensions.

4. Dark Energy: In the 4D Time model, if 4D Space is expanding, then, because of the reciprocal relationship, 4D Time must be contracting. ZPE has the necessary attributes to be the energy that manifests as an expansion of the universe, but only a small portion of ZPF is necessary to explain this expansion. This expansion of the universe is attributed to Dark Energy. This correlation means that the contracting energy of 4D Time is the Dark Energy that is driving the expansion in 4D Space. The current model of the Universe states that it consists of approximately 70% Dark Energy, around 25% Dark Matter, and the leftover five percent is ordinary visible matter. Most of our matter in the observable universe is not in 4D Space, but looks like it might be in unobservable 4D Time.
5. Balance: In the 4D Time - 4D Space spacetime model, a balance in the universe is maintained because of the reciprocal relationship between space and time. The ZPF also has a balance mechanism with prevents the destruction of observable, and unobservable, matter.

8 Spin

Forward, forward let us range,
Let the great world spin forever down the ringing grooves of change.

Lord Alfred Tennyson

8.1 A universe in constant rotation

In a scientific paper, Professor Michael Longo wrote about research done on the spin of galaxies in the universe. Of the result, he said:

This has led to strong evidence for a cosmic parity violation in the Universe, as indicated by a statistically significant excess of left-handed spiral galaxies toward the North Galactic Pole and an excess of right-handed in the opposite direction. This also suggests that our Universe has a preferred axis and a net angular momentum. Since angular momentum is conserved, this means the Universe must have been born spinning. We can't see outside of our Universe, so we'd have to assume it is spinning relative to other universes in a higher dimensional space. Presumably, the Big Bang was spinning initially, and as it expanded, the net angular momentum was dissipated among the galaxies. Now we still see it through the preferred spin direction. [1]

It seems that the spin we see out in the cosmos, the planets, stars, and galaxies was there from the beginning.

Spin is also a property for the smallest parts of our material universe. With the development of Quantum Physics came a property associated with particles named spin, quantized as everything in the quantum world is.

The Stern Gerlach experiment that separated a beam of electrons based on their quantum spin orientation, open up a new discussion. What part of the electron is spinning? One of the advantages of Quantum Theory is that it allows the calculation of behavior without having an accurate understanding of what is physically happening in the atomic world. So a model based on 'as if' the electron, or electron cloud, is spinning works fine. If the particle that has the property of spin also has the property of charge, then the combination of charge and spin of the particle creates a magnetic moment. As we will see, the different spin allowable for electrons gives atoms, molecules, and materials their magnetic properties.

But spin is more than a measurement of some aspect of the particle spinning. The amount of spin radically changes the way the particle behaves with other particles and with other matter, as we will discuss next.

8.2 What is spinning?

Spin is another property of particles like charge and mass. If a particle has charge associated with it, when this particle moves, it creates a current. Whenever you have a current, a magnetic field is automatically created. This magnetic field is called the magnetic moment associated with the moving charged particle. If the particle does not have charge, like a neutron or a on, then the particle still has the property of spin, .. does not generate a magnetic field.

The word particle is used to describe an agglomeration of charge or a very local electric field, and this local electric field manifests itself in matter with the properties that can be associated to a particle, such as an electron, proton, or neutron. As discussed before, Quantum Physics makes it clear that energy can be represented by either a wave or a particle. So if you speak of an electron, the term electron cloud is really more accurate, but it is more convenient to think of them as particles. So if the electron is represented as a cloud, what is really spinning? Truthfully, nobody really knows.

Quantum Physics also limits the values for spin to discrete values. Just like energy, spin can have values of 0, +/- ½, +/-1, +/-2, and so on. The magnetic moment associated with the spin can be used to separate particles with different spin. So even though physics does not understand what is spinning, the mathematics allows scientists and engineers to design hard drives that store information based on spin, and there is a whole new format for electronics, called spintronics, which uses the state of spin to store and move information through electronic circuitry.

8.3 Types of spin

There are multiple types of spin. There is the spin of the particle around its own axis, which is referred to as axial spin, or spin angular momentum.

Then there is additional spin as an electron moves in its orbit around the nucleus. This is known as orbital spin or orbital angular momentum, and this spin also generates a magnetic field.

The total amount of spin is a combination of orbital spin and axial spin, and both contribute to the magnetic properties of materials. Now, a particle—for example, an electron or a proton—can flip between the allowable spin states if the appropriate energy from an external magnetic field is given to this particle. As the particle returns to its original state, it returns in a manner that is called precession, a spiraling motion. The explanation as to why the spin axis moves in a spiral is shown in the physics experiment where you take a wheel from a bicycle, and mount two handles on the axis so that you can hold the spinning wheel. When you move the axis of the spinning wheel, your hands are twisted. The force of this twisting, called torque, is responsible for the precession. The frequency of this precession of the spin state, known as Lamour Frequency, is used in MRI imaging and Nuclear Magnetic Resonance Spectroscopy to image the body structure in medicine as well as research biology and matter to a very precise degree.

8.4 Classes of spin

It is in the classes of spin that the properties of spin can get a bit bizarre. First, if the tires on your car were limited to the rules of Quantum Physics, they could only rotate at specific rotations. As you speed up, your speed would have to jump to the next level by a quantum interval. But it is more than just having the rotation of your car tires quantized to specific rotations. With spin, there is an additional property that really sets it apart. With the new property, it is as if the tires on your car are one thing at a certain rotation, and at another rotation they behave completely differently, not even closely resembling the properties associated with the tires they were before.

The amount of spin that a particle has dramatically affected the properties that the particle has. There are two classes that sort the spin of particles.

The first class is called Fermions, and these are particles with a spin that is a multiple of a half. Neutrons, protons, and electrons are examples of particles that have spin that is multiples of a half.

The second class is called Bosons, and these particles have spin in multiples of one, or integer spins. Photons, more commonly known as particles of light, can be of any electromagnetic frequency and are an example of particles that have spin-1.

Particles with spin-½ are forced to obey the Pauli Exclusion Principle, which states that only one particle with the same spin can occupy each available energy level. So, typically, an electron with spin +½ and an electron with spin -½ pair up, and this is why all the electron orbitals have a maximum number of electrons that is always a multiple of two. This Exclusion property causes the number of electrons in each portion of 4D Space to be small. If it were not for this Pauli Exclusion Principle associated with the spin-½ particles, atoms would collapse, since there would be no limitation to how small a space all these building blocks of matter must occupy.

Particles with spin-½ are known as matter particles, since these particles make up the matter of the universe. Generally, electrons are disturbed at very low concentration, and so many examples have been used to stress that most of the space that an atom occupies is 99% empty. The reason for this is that this space is required for the electrons to move around their respective orbitals, with the actual electrons only taking up a tiny portion of this space. This property has led to fermions being labeled as the anti-social particles.

Particles with integer spin, that is 0,1,2, etc., do not have to obey the Pauli Exclusion Principle, and so you can have as many particles with integer spin in each portion of spacetime. The property has led bosons to be labeled the social particles. You can pack as many bosons into a location as space as you want to. Particles with integer spin are mediators of all the forces found in our known universe, and so are known are force particles.

If an electron (time forward) with spin +½ is paired with a positron with spin -½ (antimatter electron or time-reverse electron), then the two combine to form a photon of spin equals1 and now do not have to obey the Pauli Exclusion Principle.

The behavior for fermions is radically different than for bosons. To frame this in another way, it is as if you walk one or two mph in a circle, you and those also walking one or two mph in this same circle behave in one way. As many of you as possible can walk together. If you walk ½ or 3/2 mph, then you and those around you also walking at ½ or 3/2 mph are limited to only one partner. This behavior of spin is unlike any other property associated with matter.

The last unusual property is the amount of rotation is takes particles to come back to the same place. For particles with spin-1, it takes 360 degrees. With particles with spin-2, it takes 180 degrees. For particles with spin-½, the fermions, it takes 720 degrees. The first 360-degree rotation results in some asymmetry from the initial state. It takes another 360-degree rotation to bring it back into symmetry with its initial state. Let's explore this some more, since two times 360-degree rotation could be important in light of the extra dimensions of time.

8.5 SPIN coupling of Space and Time

In a previous chapter, the discussion focused on how velocity and mass are couplers between 4D Space and 4D Time.

Now, if we take spin, we find that it has a correlation to all three of these couplers: velocity, momentum, and mass.

1. For velocity, spin is measured in radians/sec, a velocity.
2. Particles with spin have momentum.
3. For mass, particles of spin-½ are the basic building blocks of matter. Particles with integer spin are the field particles.

So motion, in the form of spin couples space and time together.

8.6 Fermions and 720-degree rotation.

The best description I have read about rotation symmetry and spin is by Jessamyn Fairfield in this blog posting [2]. The discussion here is how rotation symmetry is different for particles with different spins. She states:

One major difference is in the behavior under rotation. When we try to calculate how rotation affects a particle with spin-0, we find that it doesn't matter: the particle is indistinguishable before and after any rotation. However, a spin-1 particle requires a 360° rotation to return to its initial state, and a spin-2 particle requires a 180° rotation to return to its initial state. This may seem strange, but what it means is that the spin value describes the symmetry of the particle. If you imagine a deck of cards, the spin-2 particles are like face cards that look the same when rotated 180°. Spin-1 particles are like number cards, which must be rotated 360° to look the same as they did when they started.

There are no playing cards that must be rotated 720° in order to look the same, and yet this is the case with spin-1/2 particles. There are few macroscopic objects that can demonstrate this property, but one of them is your hand! Place any object on your hand, palm up, and rotate it without dropping your palm. After 360°, you will find your arm to be pretty contorted, but after 720° of rotation, your arm has regained its initial position!

8.7 Conservation of Spin

Tom Bearden states this unusual behavior of rotation for fermions is the basis for how spin might be associated with energy that manifest in 3D Space and, in his theory, the familiar 1D time. In his book, *Energy from the Vacuum*, he states the following [3]:

. . . the EM energy flow "from" or "connected with" a charge, is intimately connected with the spin of the charge particle, since the charged particle spins in both the time domain and in 3-space.

His analogy involves regular time, not 3D Time, and 3-space refers to 3D Space. But his analogy, I strongly believe, applies to vector time as well. A few pages later, he continues [4]:

With a little oversimplification, a charge may be said to spin 720 degrees in one complete rotation. For our purposes, it spins 360 degrees in the imaginary plane (over ict) and then spins 360 degrees in the real plane (in 3-space). Thus the negative charge can absorb the incoming EM energy flow in the complex plane, transude or "flip" the absorbed EM energy into 3-space to begin its 360 degrees in 3-space, and the excitation can decay during that 3-space spin part of the cycle.

[Note: EM is short for electromagnetic]

In this statement, Tom Bearden has the space dimensions in the real plane and the single time dimension in the imaginary plane. For those not familiar with these terms, it simply means that space and time are in different dimensions.

In the Reciprocal System Theory using 3D Time, the three dimensions of time are compressed into one dimension of time, scalar time. So the same principle he has outlined applies to three-dimensional time as well. Where it differs is that 3D Time can also have vector spin, just as three-dimensional space has. The addition of vector spin in both will add many new degrees of freedom.

But to be clear, the interface between space and time is always one-dimensional, so this 720 degree of one-dimensional spin has all the right properties for the interchange of energy between space and time.

We discussed the flow of electromagnetic energy between 4D Space and 4D Time in terms of charge. Here, Bearden correlates this flow of electromagnetic energy to the 720-degree spin of the charge. He continues with his analogy:

The negative charge absorbs a little bit of positive time energy, transduces it into a much larger amount of 3-space excitation energy, and re-emits it in 3-space as an observable, real flow of EM energy radially outward in all directions. The positive charge receives the large amount of 3-spatial EM energy and absorbs it, transduces it back into time-energy (highly compressed energy) and re-emits it in the time domain as a little bit of highly compressed time energy.

Conversely, we may consider the positive charge absorbs a little bit of negative time energy, transduces it into a much larger amount of negative 3-space excitation energy, and re-emits this negative 3-space energy in all directions in 3-space.[5]

Bearden has described how fermions, the electron and proton, the basic building blocks of matter, can be couplers via their spin and charge. These two properties result in sources and sinks of electromagnetic energy between 4D Space and 4D Time.

8.8 Spacetime and Spin

The major couplers between 4D Space and 4D Time, velocity, momentum, and mass are all correlated to spin. The major building blocks of the material universe are fermions, that is, electrons and protons. These fermions require 720 degrees of spin to return to symmetry. In addition, the proton and electron, as fundamental particles of matter, have positive and negative charge. These charges, along with spin-½ , provide a mechanism for energy to flow to and from 4D Space and 4D Time.

In atomic physics, the spin energy is the spin state of an atomic particle measured by Nuclear Magnetic

Resonance Spectroscopy, or the Magnetic Resonance Imaging [MRI]. In 4D Space, these energies are the smallest of the atomic energies. The energy levels of atoms and molecules in order of energy in 4D Space are electronic, vibrational, rotational, and spin. Electronic energy is the difference between the quantum energy levels allowable in the atom and is commonly measured with Atomic Absorption or Atomic Emission Spectroscopy. Vibrational and rotational energy is allowable motions of molecule configurations that is commonly measured using Raman or Infrared Spectroscopy.

If the energies of 4D Space and 4D Time are reciprocal, as discussed previously, then in 4D Time, the order of energy in atoms would be spin, rotational, vibrational, and then electronic. With the energy in time being 372,000 times higher than the energy in space in the configuration of our observable universe, then spin is the dominant energetic phenomenon in multidimensional time.

Spin is motion, and motion has been established to be more fundamental than spacetime, so motion in the form of spin as we currently define and understand it, is the basis of our material universe in eight-dimensional spacetime.

9 Consciousness and Quantum Theory.

Uncertainty and expectation are the joys of life.

William Congrese

In Quantum Physics, as discussed in the beginning of the book, all of matter is described as having a particle nature and a wave nature. The particle nature of matter in 3D Space is what we typically see and measure. The wave nature of matter in 1D Time is connected to all particles in the universe. Even though we have separate bodies, we move in the same field of time. It is the same time for you and for me. So time is everywhere.

From the perspective of 3D Space, it is hard to visualize this interconnectedness over all Space, but experiments based on Bell's theorem show that this is true. Because of the quantum nature of atomic particles, *the theory cannot predict how one particle will behave, but is very accurate in predicting how an ensemble of particles will behave.*

If we consider humans, we have a particle nature, which is our bodies. We are easily identified by these bodies, and it is our primary interface into the world. We make assessments of people based on their bodily appearance.

Our personalities are crucial to our interfacing to the world, but in a completely different way. The personality of a person also affects the way that people make assessments of us. We can make physical assessments very quickly, but it can take a period of weeks, months, or years to truly 'know' a person, since it takes responses to different situations to really see all the facets of their personality.

Even though we have simple categories for the personalities of people, for example, extrovert, introvert, Type A, Type B, etc., the complete consciousness of a person is difficult to assess. To compensate for the elusive categories of personality and behavior, psychologists have designed elaborate psychology tests to quantify our conscious nature. The description of our attributes are definitely susceptible to the observer effect. The identity of the person asking the questions can change the answers. Knowing who will read the answers can change the way the answer is revealed. A trusted friend will get a different answer than a distrusted colleague, or a probing psychologist. These psychological tests purposely have multiple questions designed to help overcome our ability to reveal what our strengths, weaknesses, and overall attributes are.

All the data from these questions in the psychological test, or tests, are compiled together and correlate to derive personality traits. From this data, psychologists are good at predicting how individuals with these particular traits *tend* to behave, but they cannot accurately predict how any one individual with these traits *will* behave. The same is true for sociologists. They study the traits of an *ensemble* of people and use this data to predict how this ensemble of people will behave, but cannot predict how any one person in the society *will* behave.

We also have the ability to self-interact through our own self-talk. Sometimes people do this verbally, but generally, it is a private internal affair. We interface with this 'I' all during our awake time, but where is this 'I' located?

9.1 Where is consciousness?

A simple question like 'Where are you?' can lead to some insights. If it is asked in reference to where are you in Space, it is simple. An answer like 'I am in the kitchen' is easily understood.

But ask the same question, but with a twist, 'Where are you in your body?' People generally tend to point to their chest. But why not their head or their belly? Why do so many of us instinctively point to our chests?

We could more easily make the case that who we are is associated more with our heads, since we do all our thinking and processing with our brains.

Certainly, the heart is in our chest, which we usually associate with our feelings. Is who we are more easily identifiable as a feeling consciousness, versus a thinking consciousness? Which is more important, and why?

If our consciousness is embedded holographically throughout our bodies, it would make it difficult to point to a specific place. Does every cell contain information about who we are and the trillions of them make up all the information of us in physical form? Even with all the advancements in science, these questions still plague us.

One model is that the body represents the particle nature and this consciousness, the difficult to define 'I' is the wave nature of ourselves. This idea is not new, but what is new is the model presented in this book correlating time to consciousness.

Larry Dossey, in his book *Recovering The Soul - A Scientific and Spiritual Search*, has discussed the concept of a 'nonlocal mind,' and Dean Radin in *Entangled Minds* has used the concept of

Entanglement to present this same interconnectedness at the level of the mind.

To better understand how we as conscious humans can be part of both particle and wave domains, let's look at a few attributes of ourselves.

9.2 Memories

A good place to start this discussion is with our current understanding of biological memory. We typically think of memory as being stored in different parts of the brain. This is true for some of the memories we have, but does not account for all the memories. There is one study in which a rat was trained to negotiate a maze—then the researchers removed larger and larger portions of the rat's brain. Its ability to negotiate the maze slowly declined, but the rat never lost all of its memory. The details just eroded with each surgical removal, in much the same way a hologram loses its detail as more and more of the holographic print is cut off.

In studies of individuals who were involved in accidents where portions of their brains were lost due to trauma, it was found that the individuals usually didn't lose complete memories of portions of their lives. They typically lost particular details of the memories. These researchers believe this supports the concept of a holographic model for memory, in which information stored is distributed over the entire brain. Just as each portion of the hologram contains all the information, the details become less intense as portions of the hologram become unusable or are removed. [1]

But this doesn't explain all human memory. The holographic model contradicts neurological studies, where specific portions of the brain were stimulated with electric probes while the research subjects were awake. The researchers found that by stimulating precise areas of the brain, specific physical responses and memory responses could be triggered. So it appears that our memory must have both local and distributed or holographic properties.

9.3 Holograms

We are all familiar with the three-dimensional properties of holographic images and the attribute that the information is distributed throughout the whole holographic negative. Because of this distributed property, cutting out a portion of the holographic image does not destroy the image; it merely changes the intensity of some of the details in the image.

Holographic information is encoded in the negative, or crystal, as '*changes in speed of light*.' These changes in the speed of light change the phase of the light information, which is a key component in how we see the information as three-dimensional. An interesting coincidence.

The holographic image of a few items on a table, shown below, is our familiar 3D Space environment.

Particle-like form – Glasses on the table

Images from Optics by Hecht.

This image is easily understood. Now the exact same information is embedded in the image on the next page as an interference, or wave pattern. There is no recognizable structure since it is in a form we are not familiar with. The image looks like a random wavelike pattern. This example highlights how the same information can be in two forms, a particle-like form and a wavelike form.

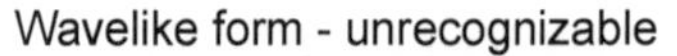
Wavelike form - unrecognizable

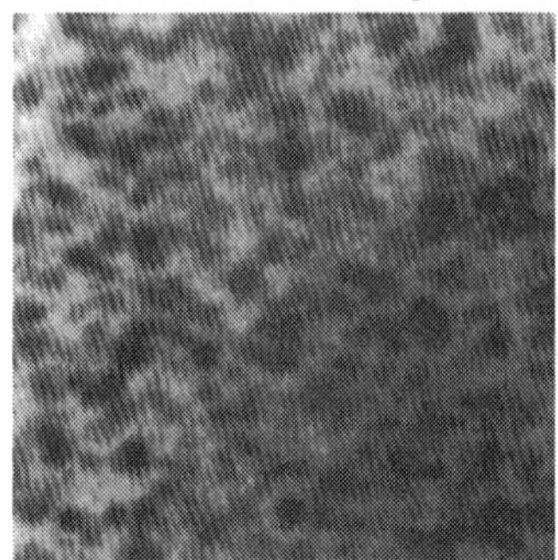

Images from Optics by Hecht.

In this example, I will make the argument that the information can be used as a model of us as conscious humans. The projected image, the glasses and pitcher in the previous image, represents our familiar and recognizable bodies. The interference wavelike pattern is our consciousness. From a space point of view, this wavelike information is not recognizable and appears distributed. But with the proper techniques, this distributed wavelike information can be extracted.

In holograms, many 3D images can be stored in the same negative by using different angles for the laser beam. When reading out the holographic information, when the angle associated with the recording of the object is used, just that image associated with that object and laser beam angle appears in three-dimensional form. So it could be with our memory. It is known that we store information based on attention and emotion. Trigger the same attention and/or emotion at a future date, and those same memories are experienced.

Another example of the same information in two different forms is shown below. In the top image, Mickey is shown in a format that is very familiar in 3D Space.

Mickey Mouse with definable structure

Atlas of optical transforms, Harburn, Taylor, Welberry

Now, using a Fourier Transform technique, where signals and images are reduced to frequencies, the image is shown after this technique below.

Mickey mouse with unfamiliar structure

Atlas of optical transforms, Harburn, Taylor, Welberry

It looks like a blob of light. Same information, but very different format. Who would have guessed what lay inside that blob of light?

The world around us can be looked at in terms of particle [space] structures or wave [time] structures, just as our memories have specific space structure forms, but also wave structures to them as well.

Could our perception of the world be completely different than what our senses are telling us? Our ears are very good at breaking down sound signals into the frequency components, in other words, doing a Fourier Transform on the sound wave. But what about our vision? The next section will look into that topic.

9.4 What do we see?

Researchers wanted to investigate how the brain processes images. The standard model was that different parts of the brain interpret different geometric shapes that we see. This model can be associated with the particle-like view of nature and 3D Space.

To investigate the electrical activity in the brain during vision, the researchers hooked up a monkey's brain to an Electroencephalography, or EEG machine. They then showed the monkey simple shapes, like the letter A, shown on the following page.

The researchers took the EEG data and correlated it to the standard understanding for processing images based on shape. They then took the Fourier Transform of the images. The Fourier transform of the letter A is shown below the letter A. They correlated the two images with the EEG data.[2]

What they found is that the correlation of brain activity in terms of the spatial frequencies of the letter A in the Fourier image was higher than just the brain's response to geometric shape.

Letter A as we know it

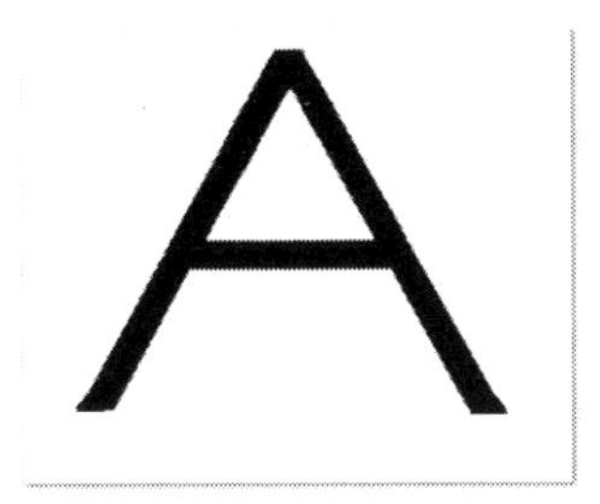

Letter A after Fourier Transform

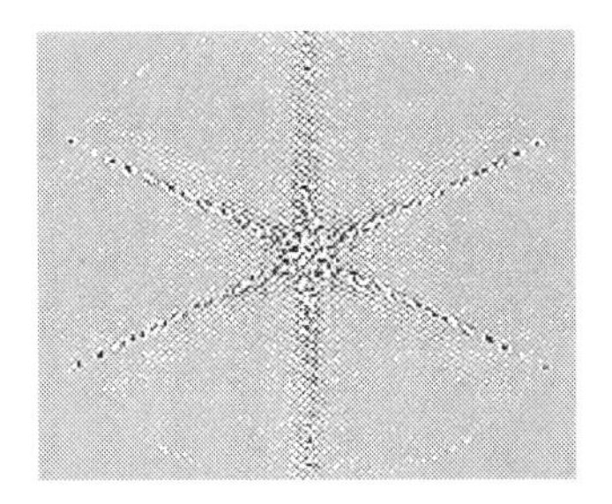

While the monkey's brain was doing geometric interpretations, it was also interpreting frequency content. It appears our brain might also be interpreting reality in terms of frequency. Frequency, a 1 / time variable, can be associated with the wavelike nonlocal nature of 1D Time and the particle-like local 3D Time. Remember, we were able to represent mass in terms of frequencies in a previous chapter, so matter can be viewed in terms of frequency as well.

These studies in memory and vision, along with our understanding of hearing, supports the idea that we sense both the wave and particle duality, or the space and time parts associated with our universe.

9.5 Self-interaction

How we interpret the world changes how our body responds to the world. We are constantly making assessments of the world around us, both from a particle and wave point of view. I believe the phenomena around placebos highlights this effect.

Placebos are usually a sugar pill given to a patient in clinical trials by pharmaceutical companies to determine the effectiveness of their drugs. Those who receive the placebo are called the control group. The objective is to determine the effectiveness of the chemistry in the pill relative to those who did not receive the pill.

These tests are conducted as double-blind tests, so that even the researchers don't know which pill is the placebo and which is the drug being tested.

Why is it important to do a double-blind test? In a study, researchers want to limit any influence the participants in the study might pick up, either consciously or otherwise from those who are giving out the pills, which might lead the patients to make assessments about what pill they're getting.

The particle nature would be the color, shape, size, taste, and other physical properties associated with the pill. From previous studies, it is known that these differences do trigger different responses in people, based on their preconceived notions and past experiences regarding these physical properties. For instance, larger pills generated preconceived notions that the pill will be more effective.

Then there is the wave nature. This would be what the doctors, or nurses, know about the pills they are administering but are not communicating intentionally. But this knowing, or belief about these pills, is communicated in subtle ways by tone of voice, body language, eye movements, and so on. If the individuals know which pill is the placebo and which is not, that knowing will change the communication in subtle ways which would then affect the outcome of these experiments.

These interactions have both of these components, particle-like associated with space and wavelike associated with time. The way a person administering the placebo is dressed and the manner in which that person speaks changes the way the patient interprets their potential authority. For example, dressing and acting like a doctor, nurse, or any other medical professional makes the placebo more effective. The tone of voice and the environment have all been found to make a difference. So it's not as simple as just a chemical and biological interaction between placebo and participant.

To add more dimensions to the experiment, not every person reacts to the same situations in the same way. Each person has their own self-interaction with these situations, and it can radically alter what people believe and how they behave.

These self-interactions, manifesting as the internal attitude that the participant brings to the interaction, is important. The more motivated they are for a recovery or healing, the more effective the placebo is. In some cases, the effect of a placebo has been staggering. Patients suffering from severe physical, mental, and emotional maladies have shown remarkable recoveries when placebos are administered to patients whose *belief or trust* in the placebo and the person administering the placebo is strong. Side effects have also been reported by participants receiving the placebo. If the participant is told their placebo is a stimulant, it has that result; if it's labeled a relaxant, it has that effect. Same placebo pill. [3]

This same interaction also applies to pain. Larry Dossey, in his book *Space, Time, and Medicine*, describes how people have rid themselves of chronic physical conditions through the practice of meditation. He labels chronic physical conditions like heart disease, high blood pressure, high level of hormone secretions, muscle tension, and a number of other conditions as 'time sickness,' since it comes from our sense of internal and external urgency against time. By finding an outlet, be it meditation or fishing, or any other activity that reduces this sense of racing time, our bodies respond by reducing the tension from our dysfunctional relationship to time.

Even the simple act of being insulted, is a self interaction phenomenon. Although it might not seem that way, when an individual says something to you, often faster than your consciousness can process it, your subconscious has made the decision as to whether this communication is labeled an insult or not. Often, the same statement given by different individuals can result in very different responses from you. The key here is that your body is waiting for a label on the statement that has been processed by your ears. Before it is labeled, it does not respond. Once the label has been given, that is, it is an insult, then a physical process is triggered. Your heart rate increases, muscles tense, your breathing changes, and so on. But this response is after a label has been given, either very quickly by your subconscious or more slowly by your conscious.

For these studies, it can be seen that our biological bodies in space are dependent on and linked to the interpretation made by our own consciousness.

9.6 Mind AND biological consciousness

Today, there is a lively discussion about whether consciousness is biologically based or if there is something much more. The discussion seems to push for an *either/or* result. Another approach could be a *both/and* result to this particular discussion. That is, consciousness is *both* a result of non-biological phenomena *and* biological phenomena. To explain this better, I will discuss an analogy that uses a car and a driver.

Let's assume we have a vehicle, and it represents a physical body. It has an electrical system, like a nervous system; it consumes energy to run; it has a computer system to control functionality, like a brain; and it's otherwise self-sufficient, once it's fueled, much like a physical body. With the necessary input, it can move or stay still. Now, let's assume the driver of the car can't be seen.

A person watching this vehicle would notice that this car would move at times, and at other times remain stationary. This observer, being of a technical and inquisitive nature, decides to probe this vehicle when the vehicle moves or comes to a stop. With his diagnostic equipment, he monitors the fluid and mechanical systems connected to the wheels as the car speeds up. All this activity causes the vehicle to move, either forward or backward, depending on the state of the gear system, as well as faster or slower, depending on the rate at which the mechanical parts move up and down.

Clearly, changes in these systems cause the vehicle to come alive, to perform actions. It makes decisions, as seen in all the electronic circuits that actively monitor all of the systems simultaneously and send feedback to control them, making the whole system behave better. All these actions could be called the consciousness of the vehicle. But even though the pattern of activities causes action, none of the systems in the car can initialize purposeful action.

The observer comes to the conclusion that the purpose of all these actions result from a higher decision-making ability. It is the unobservable driver of this vehicle that provides the intention and purpose for all the necessary and complex activity in these systems.

Potentially, intention and purpose can be determined from the data, but only after the action, not before. What if there's no purpose and the choice is to go for a random afternoon drive? Without previous pattern, no purpose can be inferred. Who is in charge of that decision? The unseen driver is.

It's a combination of driver and vehicle. That's a *both/and* process. The driver has the intention and purpose, *and* the vehicle has the ability to carry out these intentions and purposes. Both are necessary. For the driver to activate the car system, there has to be the necessary coupling mechanism between the driver and the vehicle.

To finish this analogy, the vehicle *does* impose its own set of limits. Maybe the vehicle can't do off-roading, or is too small for more than two people, or isn't running reliably enough to make long trips, and so on. The expectation would be similar for a human being, where some body structure will impose certain limits. In order for consciousness to interact with the human body in such a way that it could perform the necessary functions, functions not just of the body, but transcendental functions such as creative thoughts, I would expect there to be some physical mechanism that would be able to provide input or physical feedback when these functions are being performed. In creative people, the expectation would be for scientists to find creative tendencies in their DNA and those who exhibit tendencies toward mysticism, the expectation would also be to find these same tendencies in their DNA as well.

Now it's a bit of a chicken and egg scenario. Are these features present in their DNA because they were so inclined before birth, or is this attribute in the DNA of the family tree? Is the presence of this DNA the only reason that they are creative?

A good question is, if the driver of the vehicle in this example is related to a mind, soul, or spiritual aspect of a person, and generates the purpose and intention, then why haven't we measured or detected any presence of it yet? If the consciousness of humans, which have been shown to have wave nature, which is also associated with time, then this is understandable, since we still do not measure time directly, and therefore cannot measure consciousness directly.

10 Consciousness and Time

Problems cannot be solved in the same consciousness that created them.

Albert Einstein

This chapter deviates from Dewey Larson's idea of the meaning of 3D Time. He states:

Perhaps it might be well to point out that the additional dimensions of time have no metaphysical significance . . .. The three dimensions of time have the same physical significance as the three dimensions of space.

While I have embraced his ideas on multi-dimensional time and velocity as a coupler, I could not disagree more with him on this point. In the previous chapter, the idea that we as humans are two part beings, one part that is particle-like and another that is wavelike, was introduced. In previous chapters, the wavelike nature is correlated to time.

In this chapter, the objective is to look at consciousness and the properties that have been established for 3D Time and see how they correlate.

In Space, *coming from* and *going to* are symmetrical extensions of a location in Space called *here*.

In time, *have done* and *will do* are symmetrical extensions of time from a location in time called *now*.

We typically refer to ourselves as human beings who live *in* space who deal with a phenomenon called time. We do not think of ourselves as beings who live *in* time. But in this chapter, the idea of humans as beings who not only live *in* space but also *in* time will be explored.

The *'in space called here'* part is easy to understand. We can talk to others standing next to us, and we can relate to this 'in space' experience. But the *'in time called now'* is more difficult to grasp. If we ask people standing right next to us to describe their 'in time' experience, it could be vastly different. Some will be interested in the events around them, others complaining about experiencing boredom, still others impatient to move to the next 'in time' experience, while still others pine for a recent 'in time' experience from the past. All this is happening in the exact 'in space' location and the same moment of time, but clearly not the same 'in time' moment. So how do we quantify this 'in time' moment?

Using the ideas of Quantum Theory, the local particle-like nature is associated with our body and the nonlocal wavelike nature with our consciousness.

So far in this book, the hypothesis is that 3D Space and 3D Time are associated with the particle-like nature and 1D Time and 1D Space with the wavelike nature. The theory is that with regards to humans, the body is more associated with space, and consciousness is more associated with time. To be clear, we live both *In Time* and *In Space*, but as we will see, the implications of primarily living in space versus primarily living in time is radically different.

The implications that our consciousness in 4D Space is primarily associated with the wavelike nature of time, where, typically, contact is not necessary, is dramatic. A good place to start this discussion is to refer back to Chapter 2, where the properties of entanglement were discussed. One of these properties is that there is no limitation with regards to distance when it comes to entanglement. In regards to entanglement and consciousness, Dr. Radin asked this question in his book, *Entangled Minds:* [1]

Maybe the universe was entangled from the first few nanoseconds after the Big Bang, but how could it have remained entangled for billions of years afterward?

Einstein's Special Theory of Relativity proposed that matter and energy are different aspects of the same substance, and the atomic bombs confirmed that proposal. Thus, entanglement is a property of both matter (as in atoms) and energy (as in photons). This means that the bioelectromagnetic fields around our bodies are entangled with electromagnetic fields in the local environment and with photons arriving from distant stars. The brain's electromagnetic fields are entangled with the rest of the universe, not because of direct contact in the sense of billiard balls colliding, but because its fields interpenetrate with the energetic fields of everything else. This is also how the universe remains entangled.

10.1 Multidimensional humans

If we associate our body with particle-like 3D Space and consciousness correlates well with wavelike 1D Time, then what are the implications.

The previous sections detail that 3D Space compresses into 1D Time and 1D Time expands into 3D Time. This means that our wavelike consciousness that is in 1D Time now has shape and location in 3D Time and that our body structure that does have a location or shape in 3D Space will not in 1D Space. So all our thoughts, feelings, intentions, both conscious and subconscious, will all have structure in 3D Time.

Table 1 in Section 7.2 shows the structure of a universe that oscillates between 4D Space and 4D Time. If the universe is oscillating between these two states, and we are living *in* Space and *in* Time, then we are oscillating between these two states. So in 4D Space, our body is a local particle-like object in 3D Space and our consciousness is a nonlocal wavelike structure in 1D Time. Our body can be 'in the kitchen', but our consciousness is everywhere.

Then the universe switches to 4D Time. The information that is our consciousness, which is everywhere in 1D Time, is now expanded dimensionally and localized into a particle-like object in 3D Time, where as mentioned above, it has a location and shape that can be defined. So our consciousness can be in the 'time kitchen'. The information that is our body, localized in 3D Space is compressed dimensionally into 1D Space and the information of our space structures is everywhere in 4D Time. So what is local in 4D Space becomes nonlocal in 4D Time. What is nonlocal in 4D Space becomes local in 4D Time.

In 4D Space, 1D Time is the agent of change for 3D Space. In 4D Time, 1D Space is the agent of change in 3D Time.

A good question is why our biological structure, as well as our matter, is limited to only observe the 4D Space part of the universe. The implication now is that humans are 8 dimensional beings, but the strange part is we are only able to observe 4 of the 8 dimensions, that being masstime in 4D Space. Without 1D Time, we could not even see matter in 3D Space so we could not see our our bodies in 3D Space, never mind each other.

The ones who are aware of this are the Astrophysicists. They know there has to be more to the universe for it to behave the way it does. So they have come up with unobservable Dark Matter and Dark Energy to explain how it is that the universe can behave in the way that it does.

I believe this aspect of missing [dark] energy or matter, which are two sides of the same coin, is not only true in astrophysics, but also true for humans. We struggle to explain not only how our biology can form the structures it does without all the necessary observable information, but *our consciousness has no definable material structure in 4D Space science at all.*

Music is a complex mathematical relationship of time. The frequency of the notes are 1/ time, chords are ratios of these frequencies and the beats are measured in periods of time.

Music affects our emotions. A beautiful, happy or sad song can change the way we feel. We often use music to create, enhance or change the state we are currently feeling. Why music is capable of affecting our consciousness so much? Is it because it is structured in the same reality, that is time, as our consciousness.

Music has definable mathematical structure. You can write this structure down in great detail. Anyone, trained in the skill of reading music, independent of geographic culture or language, can understand and play it. So it should be that consciousness, one of the most crucial parts of our experiences as humans on planet Earth, can be defined in a similar way that music is defined. The evidence points to time as the basis from which consciousness can be defined.

Speaking of music, in an earlier section it was stated that lower frequencies have more energy in time than in space. Our brains as well as much of our biology operates at low frequencies, that is frequencies in the sound range and lower. Is that also a coincidence?

In time, resonance is one of the most powerful effects. Resonance with sound is often shown as a wine glass shattering from absorbing too much energy from the voice of a trained singer who is able to match the mechanical frequency structure of the wine glass. Sound waves will activate any material structure that is capable of vibrating at the frequency of the sound. The low tones coming from a helicopter flying over a house will cause structures, like windows or the walls, to vibrate if they are able to vibrate at that frequency. If the resonance frequency of a structure is not known, a driving frequency source can be set up and the frequency scanned until the structure starts to vibrate. This is seen in the Chladni and Cymatic patterns, where resonance can impart dynamic shapes in the receiving mechanical structure.

So if consciousness is in fact in time, both nonlocally in 4D Space and locally in 4D Time, what effect would resonance have?

10.2 Resonance – Like activates Like

With the theory that consciousness is in time, let's explore how resonance might manifest in consciousness. We all have used the term, 'I really resonated with that idea or with that person. Could there be more to this?

The most common example of resonance is when a tuning fork is activated and then placed in the proximity of other tuning forks tuned to the same frequency of the first fork. This can also work for a tuning fork tuned to octaves of the frequency of the first tuning fork.

This analogy can be used for communication. You have to find the right message [note] to communicate with a particular group, otherwise there will be no resonance to the message delivered. When there is resonance by a group of people to this message, a huge amount of energy can be activated. Notice, I did not say transferred, but said activated. The process is very different. In the first process, there needs to be enough energy to generate all the necessary change. In the second process, the energy is only used to deliver the message. Every individual then generates the necessary energy in response to the message.

The underlying message can be thought of as the carrier frequency, and the words and vocal delivery as a combination of frequencies just like a music song transmitted to a radio. The words in this book can trigger resonance, and with this resonance, activate a desire to investigate time. The energy to do this activity is generated by the reader, not transferred to the reader. During this investigation of time, the reader will find themselves resonating to certain ideas or repelled from other ideas.

In order to resonate with a particular listening audience, both the transmitter and receiver have to be on the same carrier frequency, much like a radio station. If the two parties are not on the same carrier frequency, the message will be heard, *but not listened to*. The process of hearing is a biological response to a sound vibration, but the process of listening requires conscious effort. The brain can convert hearing into listening when certain key words are detected, but there has to be the conscious activity of listening to trigger resonant energy, which can then be turned into action.

In consciousness, the attitudes and beliefs of a person could set up a fundamental carrier frequency for a particular topic in their life in 3D Time. If someone has a similar carrier frequency in their 3D Time field, through resonance, which happens unconsciously, the two people will activate each other. Remember, this information is made nonlocal 1D Time by the next switch from 4D Time to 4D Space. This activation could cause the two people to draw together, if they are at the same social gathering. Now the conscious reaction comes into play after the unconscious has activated a reaction to draw toward or to move away from a particular unconscious message. Let's take the 'draw toward' scenario in a place where people have congregated.

After the 'draw toward' has taken place unconsciously, the resulting conscious interaction can be easy and effortless or abrasive, depending on how one party presents their ideas. As an example, the same lyrics can be presented in the form of folk or opera. One is not better than the other, but people react differently to these forms of music, just as people who might have the same unconscious attitudes [carrier frequencies] have very different ideas or ways of expressing their ideas. People who are more combative will more often than not activate a combative reaction in others. People who are calm will more often than not activate that same response in others. The key is the words *more often than not*, because it is statistically likely, but not guaranteed in every encounter.

In life, there are situations where you need to assert yourself, and times when you need to step back. Combative is a more extreme form of assertion and avoidance of conflict is a more extreme form of looking for calm. Life seems to expect both, stepping forward and stepping back in a balanced way.

So far, the phenomenon of resonance has been used to construct a model of like attracting like in 3D Time. But what about opposites attracting, which we are familiar with in 3D Space?

The best analogy to show how both would work is one where you decided to start a blog focused on a political idea. You'd attract people who are aligned with your ideals, and they could just as easily say they have a resonance with your ideas and positions. But it is very likely your blog will also attract readers with the opposite viewpoints, who want to refute your ideas and positions. Both would happen at the same time.

10.3 Time Information Fields

With consciousness as the wavelike nature of us, then as stated, every person is linked to each other instantly. That means that the parts of our consciousness, our intentions, thoughts, and feelings are interlinked as well. You can see those dynamics much more clearly in families, or teams. The attitude of one person can change the states of the others instantly.

But how do these states interlink? In the previous section, the phenomenon of resonance was used to show how an attitude can activate others around us to choose to behave in particular ways.

Since we live in Space and in Time, we understand that proximity is important, even if it is not necessary for this interaction in 3D Time. The difference is important. Being interlinked in 3D Time is a fundamental property. The intensity of the linking can then be augmented by proximity in space, since, as humans, we react with more presence when we are in the moment of an event. The ability to hear another person's attitude in their voice makes a difference to the person who is listening.

Again, being energized by resonance in time has other properties associated with it. Coherence and timing are very important, so it is likely that a calm but coherent response can be much more effective than one that is very energized but less coherent. That is not to say that energized performances or statements cannot be coherent. When we hear these types of statements, we know it, and we feel it.

Reading a coherent and energized statement in a place and time removed from the initial statement is effective in generating either reaction, or action, for or against some idea.

Just like 4D Space and 4D Time has no directional coupling, here, location is not important, only the communication of the ideas. But what happens when a large amount of people are all finding themselves being energized by the same events? We only have to look at sports, or concerts, to 'feel' the difference that being energized individually or collectively has. In these events, a baseline resonance is already established, because the attendees are there, at some effort and expense, to witness an entertainer or athletes perform.

But just because everyone is resonating to the same fundamental action, like watching a sports event, their individual expression of the resonance is very different. This can also be seen in families. When looking at children, siblings born from the same genetic pool, and raised by the same parents in the same house can have radically different approaches to life. There are clearly attributes relating to the parents in differing ratios, but each sibling definitely has a distinct personality of their own. Some children seem to be more similar to aunts, uncles, grandparents than they are to their own parents.

Using the concepts of fields in time, here is how it might work: For now, I will call them Time Information Fields.

Remember, in 3D Time, your inner thoughts and feelings would have structure in time and are just as real in 3D Time as trees and houses are in 3D Space. The exact form of these time structures are unknown, but their properties can be known even if the structures are not. For example, we know the properties and functionality of trees, even if we could never see one or know its exact structure.

The first Time Information Field belongs to you. It is fed by your desires, your intentions, your thoughts, and your feelings. This Time Information Field is part of your consciousness, and this field either feeds or frustrates you based on what is stored in it.

Your Personal Time Information Field is effected by two other fields, Gender and Race Time Information Fields. These are powerful due to fact that they have had so much energy put into them over so many millennia. So when you identify with these traits, you are tapping into billions of those who are alive and who came before you.

All Time Information Fields are not the same for all people. For example, the Gender Time Information Field is heavily influenced by Race, but also other circumstances that will be discussed next.

At the time of your birth, you join another time field, this Time Information Field we can call the 'Family Time Information Field.' It has been fed for many, many generations by ancestors and current family members related to the this family. So this 4D Time Information Field has a history to it, stemming from the previous generations in the family, and so makes certain family traits easier to express, since the energy is readily available. In positive situations, this is like a wind or current working with you to get you to a place that's emotionally and mentally good for you. For each member of this clan, it will both nurture and frustrate them, for different reasons and at different times, depending on how your own Time Information Field interacts with this Family Time Information Field.

So far you have 4 major Time Information Fields influencing you. Your Personal, Gender, Race and Family. The order of importance is determined by each individual. They are interlinked and feedback on each other, so the relationship to you is very dynamic, and changes as you change your emotional and mental attitudes.

How does it expand beyond there? The local community has a Local Geographic Time Information Field generated by all those, past and present, who focus their attention on this community and shape its character. These properties are more easily available for those who live in them. You could visualize these Local Geographic Time Information Fields as creating a feeling for a local area that people instinctively sense and identify as local culture. It is not uncommon to hear people say they love to go to a particular area because it makes them feel more creative, more energized, more relaxed, and so on. Of course, the opposite is true, too. People avoid certain areas, are frustrated by them, and are happy to leave them for all sorts of reasons.

It is possible that the feeling of 'homesickness' comes from a person's Time Information Field missing what they have grown up in, primarily from a family Time Information Field perspective, but also from a geographical Time Information Field perspective. As humans, we all grow accustomed to a certain 'atmosphere', which can now be associated to fields, and when that changes, your system craves something familiar.

Layered on the Personal, Gender, Race, Family and Local Geographic Time Information Fields are informational fields specific to a nation. This National Time Information Field would grow and change with the focus of attention of its citizens. And so certain nationalities will interact more positively based on the content in these Time Information Fields. Finally, there would be a Time Information Field associated with this planet, where attributes of being a global citizen on planet Earth are stored.

If the speed of interaction between individuals and fields is at the speed of light, then any point on the globe is only separated by 65 milliseconds, which, when compared to the speed of our conscious reaction, is practically instantaneous. So all interactions in this field would be felt almost instantaneously. As mentioned before, for our unconscious reaction, this speed is quite slow, so on an unconscious level, these fields can clearly impact our thinking before we are even consciously aware of it.

Each individual would have different resonances with each of these fields. Some individuals have strong resonances with the local and national Time Information Fields, whereas others have resonances that are primarily dominated by family fields. This also changes over time as individuals experience events in space and in time that change their attitudes, and therefore, what they resonate to in time.

With consciousness as a time phenomena and having properties associated with time, like frequency and energy, how might these properties manifest in 4D Space. These effects will be evaluated next and this will give some insight into how this works in 4D Time.

10.4 Heart and Brain

The brain is known to have a field with frequencies in particular bands. These frequencies are measured with electroencephalogram [EEG]. There are five common states associated with brain activity: Delta, Theta, Alpha, Beta, and Gamma. The frequencies associated with these states are 0.5 - 4, 4 -7, 8 - 13, 14 - 40 and 40 to 100 Hertz. [2]

Now the electrical field associated with the heart is measured with an electrocardiogram [ECG]. The electrical field associated with the heart is 60 times greater than the brain field and the magnetic component of the heart field is 5000 times greater than that associated with the brain, as measured with Superconducting Quantum Interference Devices (SQUIDS) [3]. This makes the heart by far the largest electromagnetic field generator of any organ in the body.

In addition to measuring the electrical activity of the heart through the ECG measurements, another measurement called Heart Rate Variability [HRV] can be derived from these ECG measurements. The HRV is a measure of changes in the beat-to-beat heart rate. The patterns of these beat-to-beat measurements show changes between the parasympathetic and sympathetic branches of the autonomic nervous system (ANS) [4]. In positive emotional states, such as appreciation, love, or compassion, the HRV signal has a highly ordered or coherent pattern in the heart signal. For negative emotional states, such as anger, frustration, or anxiety, the HRV signal is more disordered and chaotic.

10.5 Entrainment from heat and brain fields

When EEG and ECG measurements are taken and the signals between the two processed, a signal called the Heartbeat Evoked Potential [HEP] is detected. The Heartbeat Evoked Potential is an electrical potential detected from the nervous system of a patient following the electrical signal coming from the heart.

This HEP signal is weaker than the ECG signals from the heart, so signal averaging and signal lock-in techniques are required to detect this signal. Colin McCraty is a researcher who detected that this electromagnetic portion of the HEP signal arrived at the brain instantly, whereas other neural signals are detected at least eight milliseconds later, with still other neural signals are detected continuously throughout the cardiac cycle.

During the normal activity of the heart, a pressure wave is generated throughout the arteries that is much faster than the pressure due to the flow of blood. The pulse measurement commonly taken is due to the flow of blood, this slower component. The other pressure wave reaches our brain 240 milliseconds after contraction of the heart and is correlated to alpha electrical activity in the brain. [5]

Expanding on the correlation between the heart and brain rhythms, another experiment was conducted by the same researcher. In these experiments, subjects were setup to face each other at a distance of five feet, and the subjects were asked to focus on achieving coherence in their Heart Rate Variability signals. The subjects were not told of the purpose of the experiment. While the subjects were doing this focusing activity, the researcher monitored the electrical signals of both subjects' brain and heart. It is pointed out by the researcher that it is known that the alpha rhythms of the brain can *synchronize* to external stimulus, for example, light and sound pulses, and so the intent of this experiment was to see if the heartbeat of one subject could be measured in the brain wave patterns of the other subject. What the researcher found is that the alpha rhythms of one subject's brain did synchronize to the heartbeat of the other subject. The researcher points out that at conversational distances, it is possible for the magnetic signals of the heart to influence the brain rhythms of the other person in a conversation. [6]

10.6 Coherence/entrainment.

In the previous section, the term coherence was referenced. What does this term mean? In physics, coherence is associated with an orderly correlation between electromagnetic waves. For instance, a laser is an unusual light source because the photons in the laser beam have a correlation in their phase for relatively large amounts of time. When you compare this phase correlation in time to an ordinary lightbulb, the lightbulb has an extremely short correlation period.

This property of coherence of the laser beam is what allows holograms to be recorded. As the coherence of the beam is reduced, the holographic effect is reduced because the depth of field of the hologram is reduced.

When electromagnetic waves are processed and viewed in terms of frequency patterns, the patterns have orderly peaks. With coherence signals, these signals can be locked together. Incoherent waves have a frequency pattern that looks like fields of grass, that is, disordered, with no discernible pattern.

Another definition for Coherence is: [7]

The quality of being logically integrated, consistent, and intelligible, as in a coherent argument. A related meaning is the logical, orderly, and aesthetically consistent relationship among parts. Coherence always implies correlations, connectedness, consistency, and efficient energy utilization.

In another paper by the same author, the idea of coherence is expanded. [8]

In this context, thoughts and emotional states can be considered coherent or incoherent. Importantly, however, these associations are not merely metaphorical, as different emotions are in fact associated with different degrees of coherence in the oscillatory rhythms generated by the body's various systems.

Now entrainment is when one signal impacts another. The second signal can change its properties based on the structure, or information, of the first signal.

With coherence, entrainment can be correlated. For example, entrainment is observed between heart rhythms, respiratory rhythms, and blood pressure oscillations. In experiments measuring Heartbeat Evoked Potentials, it was found that the brain's alpha activity is naturally synchronized to the cardiac cycle. This entrainment could be enhanced with a patient/person focusing on positive emotional states.

Resonance, discussed in a previous section, is also part of this effect. As stated in the paper by Rollin McCraty: [9]

When the human is operating in the coherent mode, increased synchronization occurs between the sympathetic and parasympathetic branches of the ANS, and entrainment between the heart rhythms, respiration, and blood pressure oscillations is observed. This occurs because these oscillatory subsystems are all vibrating at the resonant frequency of the system. Most models show that the resonant frequency of the human cardiovascular system is determined by the feedback loops between the heart and brain. In humans, and in many animals, this resonant frequency is approximately 0.1 Hz, which is equivalent to a ten-second period.

10.7 Other Field Influences

Alexander Tchijevsky, a Russian scientist, correlated human history to solar activity, by constructing an Index of Mass Human Excitability. He used the period between 1749 and 1922 to compile the histories of 72 countries. The results showed that 80% of historically important events occurred during the roughly five years of solar maximum. [10]

Solar maximums are associated not only with increased human aggression, but also increased human creativity. It seems that the energy of the sun is just that, energy. It is our choice as humans, individually and collectively, that polarizes this energy one way or another, turning it into coherent activity of creativity or incoherent aggressive behavior.

In Robert Becker's book, *The Body Electric* [11], he organized a study with a colleague to evaluate the response of 28,000 patient admissions in eight hospitals against 67 magnetic storms in the earth's magnetic field over a period of four years. The source of the magnetic storms were cosmic rays from the sun. They found a significant correlation to the number of admissions *days* after a magnetic storm.

When a study was setup to evaluate patients already hospitalized, the nurse reports revealed various behavioral changes in almost all the subjects one or two days after solar cosmic ray activity decreases. Magnetic storms decrease the amount of cosmic rays reaching the earth's surface. The researchers found the delay interesting since low-energy cosmic ray flares from the sun are known to produce strong disruptions in the earth's field one or two days later. [11]

In an investigation conducted by Frank Brown at Northwestern University and Rutger Wever from the Max Planck Institute, into oysters, known to open to incoming tides, it was found that oysters in an aquarium with constant light, temperature, and water level still opened and closed their shells in time to their companions at the bay. When the oysters were flown inland to a place 1,000 miles away along the shores of the Great Lakes, over the course of a few weeks, the oysters responded by shifting their tidal responses, as if they knew they had been taken 1,000 miles westward. In addition, they also seem to have suffered jet lag. [12]

In further work done by Brown to investigate animals' responses to magnetic fields, he chose a mud snail Nassarius, found everywhere in the world. He placed the snails in a uniformly illuminated box with an exit hole facing magnetic south. He documented their movements and experimented with magnets to see what effect he could have on their bearings. When he correlated their movements coming out of this exit hole, along with the magnetic experiments, Brown stated the following: [11]

It seemed as if the snail possessed two directional antennas for detecting the magnetic field direction, and that these were turning, one with a solar day rhythm and the other with a lunar day rhythm.

Brown also found that the stages of cell cycle time is correlated to roughly a day. While the chromosome activity takes only a few minutes, the activity preceding this chromosome activity, including DNA duplication, takes about one earth day, or one rotation of the earth around its axis. [13]

Rutger Wever did research on humans to determine our response to the earth's electromagnetic field. He had two underground rooms built to isolate people from the passage of time due to light, temperature, sound, and other ordinary clues. The two rooms differed in that one was shielded from the earth's magnetic field. He observed several hundred subjects, some for as long as two months. He plotted biomarkers such as urinary excretion of sodium, potassium, and calcium, body temperature, and sleep cycles. The results showed that people in both rooms developed irregular cycles. The people in the room that still had the earth's field were able to keep a circadian rhythm close to 24 hours, or sometimes a harmonic of 24 hours, that is 48 hours. But the people in the shielded room had much longer circadian rhythm and had metabolic rates with no correlation to each other. Wever then introduced tiny electromagnetic fields into this shielded room and found that only one frequency, around 10 Hz, has a significant health restorative effect on the subjects.

The results of these weak electromagnetic fields around 10 Hz have been confirmed in guinea pigs and mice. These results point to a frequency that seems to induce some entrainment in the physiological behavior in both humans and animals. Wever did not choose this signal randomly, but based it on the frequencies found in the electromagnetic field of the earth. [14]

The frequency spectrum of the earth's electromagnetic field [Schumann resonances] extends from roughly 3 Hz to 60 Hz, with peak frequencies of 7.83 (fundamental), 14.3, 20.8, 27.3 and 33.8 Hz. These frequencies overlap amazingly well with the frequencies of the brain. But correlation is not causation. Rutger Wever's work seems to show reason to believe there is causation and more work should be done in this area.The frequency spectrum of the earth's electromagnetic field is due to the fact that the earth's surface and the ionosphere form an electromagnetic resonant cavity that produces pulses in the magnetic field at extremely low frequencies when excited by lightning strikes.

Robert Becker states that no longer could we expect changes in the magnetic environment to be as clear and obvious as changes in oxygen, food supply, or temperature. In addition, the earth's field has many cyclic changes in it. The earth's magnetic field, derived from the planet's molten iron-nickel core and charged plasma gas in the ionosphere, varies with the lunar day and month, as well as changes based on our revolution around the sun.

Solar flares bring cosmic particles into the earth's field, causing magnetic storms, as previously stated. In addition to magnetic storms, the flipping of the poles of the sun during the solar cycle also affects the earth's field.

As humans and animals, we are bathed in fields that affect us in global ways. This effect links us all to the same phenomena in varied ways, but it points to the fact that we are entangled with fields of the earth, our moon (planetary bodies), and our sun (star) in ways we are just uncovering. [14]

In this discussion about fields, I have highlighted some properties, like that of coherence and entrainment, that these fields possess. Remember, these properties apply not only to Time Information Fields in the local particle-like form, but also to the nonlocal wavelike form. In this latter form, all fields of information in the universe can interact with us. The fact we do not consciously recognize this information is something we should consider. What would it take to be actively aware of this information, and what could we do with it, both individually and collectively as a human family?

10.8 Morphogenetic Field

So far, the discussion has been about time as a field of information with interactions to other people's fields, as well as fields from our own earth, moon, and sun. In a previous section the discussion was about how these Time Information Fields store information regarding psychological properties of individuals, families, communities, nations, and planetary groups. But is the information in these fields utilized for more specific tasks, like influencing the two key properties of space and time, that being form and energy? In this section the discussion will be about this exact topic.

Rupert Sheldrake, in his book *Morphic Resonance*, lays out a model for how a field, he labeled as morphogenetic, influences the way cells grow, thus impacting the biological structures of plant, animal, and human structure. But he does not limit it to just biological structure, since he discusses the impacts in behavior in a very similar manner to how I discuss how Time Information Fields impact human behavior. In the Time Information Field model, 4D Time has the ability to affect structure through the dynamics of time, or frequency. Cymatics shows how these structures might manifest from frequency. Remember, mass is dependent on the variable frequency, that is, mass equals frequency times the constants (Plank's constant / speed of light squared).

In Dr. Sheldrake's book, he discusses details of the morphogenetic field that are very closely correlated to the structure of 4D Space and 4D Time. In particular, his discussion on form and energy is relevant. He states: [15]

In the most general terms, form and energy bear an inverse relationship to each other: energy is the principle of change, but a form or structure can exist only as long as it has a certain stability and resistance to change.

This sounds just like the reciprocal [inverse] relationship between space and time, where form is the 3D Space component and change [energy] is the 1D Time component.

Dr. Sheldrake details how this relationship between form and energy is easily seen between temperature [energy - time] and matter [form - space]. As you change the temperature, the form of matter changes from solid to liquid to gas, and finally plasma, and vice versa.

But the real conflict of how shape is determined is plainly evident in the forms of biological structures. Although DNA is often quoted as the basis for biological structure, there are important facts that Dr. Sheldrake points out. These conflicts, detailed by him in his book are listed below. [16]

1. *The human genome has only about 25,000 genes. The fruit fly has 17,000, the sea urchin 26,000 and the rice plant 38,000.[intro 6%]. There is a larger difference between species of fruit fly and mice than between humans and chimpanzees.*
2. *Within the same organism, different patterns of development take place while the DNA remains the same. Consider, for example, your arms and legs: both contain identical cell types (muscle cells, connective tissue cells, etc.) with identical proteins and identical DNA. So the differences between the arms and the legs cannot be ascribed to DNA per se; they must be ascribed to pattern-determining factors that act differently in developing arms and legs. They also give rise to mirror-image patterns in right and left arms and legs. The precision of arrangement of the tissues—for example, the joining of tendons to the right parts of the bones—shows that these patterns are established in detail and with precision. The mechanistic theory of life means that these factors must be regarded.*

3. *Even if physical or chemical factors affecting the growth of an arm, the formation of an eye, or the development of an apple are identified, this raises the question of how these factors are themselves patterned in the first place.*

Because of these issues, Dr. Sheldrake states that the evolution of the organ and that of the molecule are independent. He asks, "*If genes and proteins do not explain the differences between chimpanzees and us, then what does?*"

To answer this question, he postulated a morphogenetic field. Morphogenesis is defined as the "coming-into-being of characteristic and specific form in living organisms." [17]

His proposed morphogenetic field is an attempt to deal with the many unanswered issues of how form comes into being, the regulation of cell behavior, biological regeneration, as well as the speed at which morphogenesis happens. The issue with speed is explained in a protein folding example, where for a chain of 150 amino acid residues, there are 10^{45} possible random conformations. The number is lower due to the fact that a number of these conformations are sterically impossible. But if each potential shape could be done with a molecular frequency of 10^{12} Hertz, which he states is an overestimate, it would take *10^{26} years* to generate all the possible conformations. Now contrast that to the synthesis and folding of ribonuclease or lysozyme protein that can be done in two minutes, or roughly in a *four-millionth of a year*, to reach the lowest free energy shape, and the problem becomes clear. The protein must have a way to determine the lowest free energy state so extremely fast. [18]

Dr. Sheldrake states,

This discussion leads to the general conclusion that the existing theories of physics may well be incapable of explaining the unique structures of complex molecules and crystals; they permit a range of possible minimum-energy structures to be suggested, but there is no evidence that they can account for the fact that one rather than another of these possible structures is realized. It is therefore conceivable that some factor other than energy "selects" between these possibilities and thus determines the specific structure taken up by the system. The hypothesis that will now be developed is based on the idea that this "selection" is brought about by a new type of causation, at present unrecognized by physics, through the agency of morphogenetic fields. [18]

He goes on to say that the mechanistic view of science, [19]

. . . assumes that all this can be explained in terms of self assembly. . . . This is rather like saying that a house can build itself spontaneously as long as the right building materials are delivered to the building site at the right times.

I definitely agree with this statement. Later, Dr. Sheldrake compares the architectural plan to information. The plan can be destroyed, whether the house is built or not, but it has no energy of its own. I agree that in space, this has no energy of its own. But in time, this plan, with all the energy spent by the homeowners, both in their dreams and visualization of this house and the formulative discussions with the architect, this plan is energy in unobservable time. It is in the process of becoming a reality in observable space because of the unobservable energies put into it in time.

These morphogenetic fields have the properties of the wave nature, since information in the field is accumulated not through physical contact, but through patterns of 'coming into being.' These patterns can be changed over time, but the patterns of the past continue to guide the patterns of the present and future. Change, the necessity for or the prevention of, is what time does in space. So it is within the realm of possibilities that these actions that Dr. Sheldrake associates with morphogenetic fields are in fact the manifestations of Time Informational Fields.

10.9 Morphogenetic field and physics

In correlating these morphogenetic fields to theories in physics, I will highlight two theories that Dr. Sheldrake discusses. In the first approach, these fields might be due to additional dimensions of spacetime. A theory right in line with what I am proposing here.

A second approach is related to the Quantum Vacuum. He references Quantum Theory, where all electromagnetic fields are derived from the Quantum Vacuum. Virtual photons appearing and disappearing into the Quantum Vacuum mediate the properties of these fields. Dr. Sheldrake theorizes,

Thus, all molecules within living organisms, all cell membranes, all nerve impulses, and indeed all electromagnetic and chemical processes depend on virtual photons appearing and disappearing within the all-pervading vacuum field of nature. Could morphic fields interact with regular physical and chemical processes through the vacuum field? Some theoreticians speculate that they can and do. Theories of these kinds may help to relate morphic fields and morphic resonance to the physics of the future. But at present, no one knows how the phenomena of morphogenesis are related to physics, whether conventional or unconventional. [20]

As discussed before, the Vacuum field and morphic field could both be based in the additional dimensions due to 4D Time.

10.10 Time and DNA

With fields in time potentially being the patterning information for matter, then what are the implications for DNA? The view expands to the idea that DNA is not only in multidimensional space, but multidimensional time as well. Like the conservation of charge and expansion/contraction in 4D Space and 4D Time, DNA could have a whole other set of properties in unobservable multidimensional time that clearly could affect the patterning and growth of biological matter, possibly in ways that we have just discussed.

As Dr. Sheldrake so succinctly put it, the construction of biological structure is much more than just the delivery of the right materials at the right time.

The part of the DNA in time could be the interface to the Time Information Fields that hold all the architectural plans of how biological cells organize. These architectural plans are not static, but are influenced by all the nonlocal information available everywhere in spacetime.

10.11 Human Consciousness in Time

In this section, I would like to use an analogy to highlight the implications of what it means that our consciousness is a nonlocal phenomenon.

The model postulated is that what we think and feel is projected into layered Time Information Fields that surround us. In accordance with entanglement, the interaction is not just limited to local fields on planet Earth, but with all other fields in the universe. It is a big concept to accept about our place in both 4D Space and 4D Time.

How would thoughts and feelings inside a person affect their reality in this new eight-dimensional spacetime? Let's hypothesize that there's a person who has a creative thought which generates a frequency in 4D Time. This frequency in 4D Time resonates with similar frequencies in 4D Time, manifesting what would look like an attractive force in 4D Space because the amount of energy increases with this resonance from other frequency sources in time. If other people have already paid attention to this idea, then, through resonance, these ideas can appear as new insights to this individual. If this creative idea has very little previous attention paid to it by others, then the person who is the source of this idea will have to put more effort into generating their own insights. As they put more energy into this idea, the path for someone else, independent and unknown by the first person, to gain similar insights will be easier. I specifically use the word opportunity, because if these insights are blocked, or filtered through the mental constructs of this second individual, their usefulness could be reduced. So what happens in 4D Time with this resonance is separate from how this energy of resonance is received by any one individual.

When more attention is paid by this one individual or others interested in the same idea, or from previous efforts behind this idea, this resonance generates a flux, a flow of energy. Eventually, when a critical flow is reached, this flow of energy per unit time is felt as power. Once it has reached this critical point, the energy would be self sustaining, making it much easier to realize this idea, or in modern parlance, having it go viral. Before this critical point, if insufficient flux or flow is generated, then the idea would slowly stagnate.

Now let's look at another analogy. In this second example, let's examine two people confronted with an important life decision. We have two people in their senior year of high school. The first student is Clara, who's very talented in science, knows that she wants to study neurology, but is trying to determine the right school to maximize her future career. The second student, Jeff, has no clue what he wants to study or where he wants to go, but he's talented in a number of different areas, so he's having a hard time just dealing with the thought of choosing a school and being pinned to a choice so early in his life. Both start with their potentials at the same point in time. We can take a snapshot of the present moment, a 2D slice of one point in 4D Time, much like a slice from a loaf of bread in our 4D Space. The loaf represents all of the 4D potentials in consciousness, but we are only interested in the possibilities in their life at that moment in time.

If you're interested in another particular moment, you slice the bread at that point, and you get a different 2D time slice of these potentials. When you slice the bread, the slice might have large or small holes, or a different grain structure. If you string a whole bunch of slices together, the points of potentials in each part of the 2D slices of time turn into threads or strings that weave through the bread like it might in 4D Time. Some of the 4D Time threads remain isolated, some join together and reinforce these potentials, other threads cross each other, and some threads survive while others terminate. These threads represent potentials in a person's life. In 4D Time, these are as real as doors and windows in 4D Space, but in this 4D Time, they're the hopes, dreams, ideas, creativity, aspirations, and so on that people focus their intention and attention on.

In 4D Time, we stated that all possibilities are stored as potentials, much like energy is stored in electric and magnetic fields. The more potential a possibility has, the more powerful or forceful the field that can be generated, and the higher the possibility of manifestation in 4D Space.

So each potential outcome, or possibility, is mapped into this slice of time. Now, if we're only looking at career potentials and filtering out all the other life potentials for these two individuals, here's what we would see: For Clara, we would see a few strong points clustering in this slice around one idea relating to biology and neurobiology. We might see one point on this slice relating to pathology, since Clara has a very good mind for solving complex situations, and she's played with the idea of specializing in the forensic sciences. We can imagine that a university program that has similar concepts will exert a pull on her.

When we look closer, we see that the strong points all relate to schools she's applying to. Some of the points are larger, meaning they have a stronger potential manifesting for her at that moment, as shown in the figure on the next page. This could change just a few time slices into the future, but for now, we're only examining this one-time slice, which is a day in late September of her senior year in high school. The points in this reality are a cross-section of who Clara is, the amount of energy she's put into the subject, and each school she's interested in.

Now, each school has a potential in this dimension as well, and the interaction of who Clara is and her efforts intersect with the possibilities offered by each school; and so a new potential is created. In late September, School A looks like the best fit. School B doesn't appear to fit Clara's personal potential as nicely as the others, but Clara has added a lot of energy to it, believing it to be the best fit from her perspective, and so it also manifests as a strong point or potential. If someone were to dig deeper, it might be obvious why School B manifests as less of a potential when interacting with Clara's own potentials, but that would require evaluating multiple time slices and going back in time to examine the issues between Clara and the potentials surrounding this school for her.

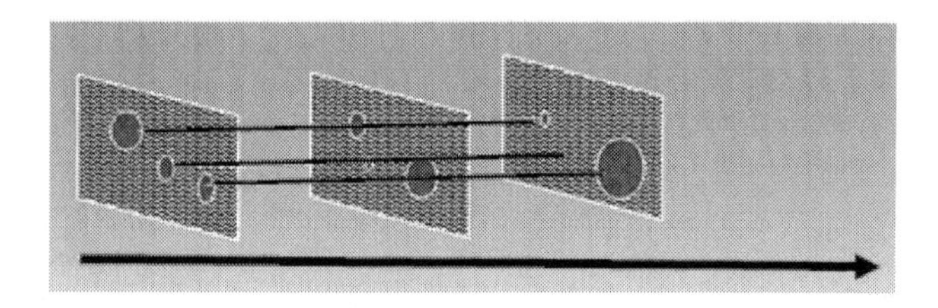

School C is another strong possibility. In this case, the school's potential has a much higher energy content for a fit with Clara—but she hasn't put as much energy into it, because with her strong, rational mind, she often overrides her intuitive side. If you were to ask Clara, she would tell you her choices are schools A, B, and C in that order. But if you were to read into the potentials, you would say A, C, and then B.

Meanwhile, for Jeff, a time slice in September would reveal many small dots all over the place. It would look a bit more chaotic, with a few correlations, but no strong ones. There would be a bunch of schools with potentials for Jeff and who he is at this moment in time. There would be clusters of points representing talents that he's put energy into and has strong feelings toward. Some of the points have links between them, because Jeff sees them as interrelated. His writing, his technical interests, his drawing, and his travel interests are all creative expressions of who he is, but if he has to choose a major, he finds himself paralyzed. Choosing one seems to mean the others might not be a part of his life in the way he envisions them.

In the 3D potential reality, we see threads of potentials that can turn into possibilities with enough attention. In our student examples, we can see how visits to colleges reinforce threads or cause them to diminish and, in some cases, to terminate. During the acceptance part, threads will continue; some grow and others will terminate. Finally, for one student, the financial aid part will bolster or diminish potentials and, eventually, a decision will be made.

It could be said that when people talk about gut feelings, these people are able to tune into these possibilities and see which ones are manifesting as a better fit for them. This could be thought of as a reflected wave from the future coming from a potential field that's manifesting due to the energy being pumped into it from the person's effort in the present regarding that future event. If the reflection is in alignment with their present ideas about the future, a good feeling results. If this reflected wave isn't in alignment, then there's a poor gut feeling.

In this example, the idea is that everything you intend, think, and feel is a dynamic playing out in time and there is feedback, nonlocally, to all this conscious activity. The feedback in time can also explain how it is we feel uncomfortable with certain decisions, in spite of the fact that in 4D Space, all the facts would lead to a decision in that direction. It is not uncommon for people to make major decisions based on their gut feeling.

The example given here highlights the fact that we *are* really spacetime beings living *in* multidimensional space and *in* multidimensional time. The results of our activities in space are immediately obvious, but we cannot see the activities in time and so have trouble correlating the results of these activities.

But it is not just that each individual has a presence in spacetime, but their feelings and thoughts about their community, state, nation, and their role as a global citizen now have the mechanisms to be real in multidimensional spacetime.

11 Balance of One.

You can never use the inside of the cup

Without the outside.

The inside and the outside go together.

They're one.

Alan Watts

11.1 Centered on One.

Let's assume that everything in the universe has a motion that is exactly the speed of light. Now we normalize the speed of light to one. Any speed slower than the speed of light is less than one, and any speed faster than the speed of light is greater than one. This one is an equal number of units of space divided by the same number of units of time, which Dewey Larson refers to as the Reference Frame of motion, a state in which the material universe has no discernible structure.

If everything moved at the speed of light, the results would be a very uniform universe. Only motion different from *one* would represent any structure at all.

Think of it this way: If the only frequency in the universe was the frequency represented by the color green, the whole universe would be one uniform color of green. We can call this color green, unity. You could not discern one part of the universe from any other part of this unity universe because of this uniformity. This uniformity is the reason for the undetectable nature of 1D Time as well as the Zero Point Field.

Now we introduce small changes. If a small change results in a slower frequency from unity, the color yellow-green appears. At places where the small changes appear faster than unity, blue-green appears. Now the universe has structure starting to appear, and areas of yellow and blue can be discerned from the overall background of unity, the color green. As larger changes from unity appear, then yellows, reds, infrared and indigo, purples, and ultraviolets appear in this universal structure.

But all changes are referenced from green. In this manner, the changes in color represent what happens when the symmetry of the universe is changed from total balance, a *Balance of One*. So a charge dipole would be a change from green, which represents total balance of all charges, to a color of yellow on one side [we could label as positive polarity] and blue on the other [negative polarity].

In a universe where all speeds are the speed of light, the natural local reference for unity, there would be no structure, only motion. All motions slower than the speed of light manifest as matter with, as in the previous example, yellow through infrared properties. All motions faster than the speed of light manifest as matter with blue through ultraviolet properties.

If we are talking about consciousness, this green represents the 'One' moment in time. This *One* moment is the same everywhere in the universe since it is nonlocal and therefore not dependent on location. The past is a universe moving *slower* than the speed of light, and the future is a universe moving *faster* than the speed of light.

Note: The name of One moment is chosen in order to prevent confusion with the now moment described in physics.

Special Relativity only allows for a common now moment for those in a common reference frame. For humans, this frame of reference is the surface of the earth, and even though velocities can change at different places, for example the equator [fastest] to the poles [slowest], the speeds are too slow to impact the simultaneity of an event. So for humans on planet Earth, the now and One moment are the same. But at very large distances, the now and One moment are not the same. The One moment is always the same since it is not dependent on location, but the now moment is dependent on location.

In this chapter, I call this point where the speed of light is normalized to one, the Balance of One. I labelled it this because it has implications not only in the material world, but in consciousness as well.

11.2 Balancing Act

In 4D Space, motion is a change in location in 3D Space with respect to 1D Time. In 4D Time, the motion is a change in location in 3D Time with respect to 1D Space. So an atom has a location not only in 3D Space, but also in 3D Time. The motion of this atom is affected not only by nonlocal 1D Time, but also by nonlocal 1D Space.

Since 1D Time is everywhere in 3D Space, and 1D Space is everywhere in 3D Time, this Balance of One, then, is expected to be everywhere in Space and Time. It has to be in order to keep this universe in total balance. The Balance of One contains the points where 1D Space divided by 1D Time has the same value as 1D Time divided by 1D Space. That is, the Balance of One, common to both 4D Space and 4D Time has the property where

velocity = 1 / velocity

which can be stated as

space / time = 1 / (space / time) = time / space

which is the speed of light.

To understand how the Balance of One point changes between 4D Space and 4D Time, I will use a nice example given by Dewey Larson of air in a balloon for an expansion and contraction analogy. [1] The air molecules in the balloon represent the Balance of One points, which in this example we will call the 4D Space matrix. If energy is put into the air, the molecules move away from each other because they are moving with more energy, so the distance between the points of the matrix increases and the matrix of points expands. This is how gases expand when heat is put into them. The molecules absorb this heat energy and vibrate over larger distances. In the process, each molecule of gas bumps the other surrounding molecules, pushing them further away. Each point in the matrix is expanding away from all the other points in the matrix. This motion is only scalar, since it moves in all directions outward at the same rate.

If a balloon is filled up outside on a very cold day with this cold air to a certain radius, then sealed and brought inside to a warm room, the balloon would expand due to the increased energy level of the air molecules due to the heat in the room. So this action represents 4D Space expanding.

In the 4D Time matrix, the matrix is contracting, since motion in time is in opposition to motion in 4D Space. In this scenario, you move the balloon from the warm room to the very cold air outside. The balloon will contract due to the outside cold air lowering the energy levels of the air molecules inside the balloon. This action represents 4D Time contracting.

In order to maintain balance in the universe, if in 4D Space all points in the matrix are expanding, then in 4D Time, all points would be contracting by an amount that is reciprocal to the expansion. Imagine two balloons connected to each other so a fixed volume of air is passed between them. If one balloon is expanding, it is taking more of the shared volume of air in the two balloon system, so the other balloon has to contract.

This model supports a balance of gravity and antigravity, or levity, between the 4D Space and 4D Time portions in our universe which is coupled through the one-dimensional, or scalar, speed of light.

So if spacetime in 4D Space is expanding, as Astrophysicists have measured, then the oppositional motion in 4D Time is a contraction, or gravity.

11.3 Consciousness and the Balance of One

Before there was spacetime, the spiritual texts hinted at a nothingness. It is possible that all these spiritual references to a nothingness is, in reality, a reference to everything in total balance. For all positive charges, there were equal negative charges. For all the gravity, there is antigravity, or levity, to balance it out. This Balance of One was just light. It all was in perfect balance and unobservable to us.

From a metaphysical point of view, *one* is seen as the number of divinity, of unity. It's part of all things and leaves all things unchanged. This is also true in math and science, since *one* doesn't change any of the other numbers when it's used in equations. In our language, becoming one with something is considered an enlightened experience.

If you take a perfect circle and you polarize it, you have an ellipse. So one point at the center of the circle is divided into two focus points. These two foci could represent duality, which could be the polarization of spacetime. One foci is a positive emanating source, while the other is a negative accepting sink. In electrical terms, the positive charge is a source, the negative one a sink. In fluidics, a spring coming out of the ground is a source and a drain is a sink. The drain is also in the sink, but you get the point. Gravity can be thought of as a sink, levity as a source. In almost all of matter, the ellipse with its dual points of polarity, is the common geometry.

As stated before, the Balance of One is where we continuously experience time as the present. On one side of the Balance of One, all velocities are slower than the speed of light, and life is always experienced as in the past. On the other side of the Balance of One, all velocities are greater than the speed of light, so life is always experienced in the future.

The way we experience our spacetime, the velocities are always slower than the speed of light. Even when we think we're in the present, as mentioned in Chapter 1, we're really in the past. It takes time to process what we see, hear, feel, and so on, and even this small amount of time puts us in the past.

To be truly in the One moment, we have to bypass our thought processes, to be in the zone. We have to link into what's been described as the super-conscious portion of ourselves. As conscious humans, not just biological entities, portions of us are connected with all these interlinked layers of reality.

The way I envision it is that super-conscious portion of ourselves is located outside of the spacetime construct, and therefore not influenced by the duality construct of space and time. It is a place of null time and null space, the place from which space and time are generated.

This super-consciousness is probably the part holding the truth about who we really are outside this spacetime construct, and offers a way to decode the duality creations and distortions built into the local space and time domains we experience. The super-conscious, like the Zero-Point Field, represents the portion of us that has almost infinite potential. But only a portion of this huge potential manifests as energy, the energy of our consciousness and our biology. In the same manner as 1D Time manifests everywhere in 3D Space and 1D Space manifests everywhere in 3D Time, our consciousness is manifest in all Space and Time.

The implication would be that there's a portion of each one of us existing throughout Space and Time in the whole universe. It seems that awakened humans try to communicate this idea that we are part of this whole cosmos, it's just that we don't consciously understand or know it yet.

I envision that the Balance of One acts like a spacetime filter. It filters the information in these Time Information Fields and prevents it from flooding into our daily consciousness, much like a filter used to process and acquire particular portions of information from a huge database. The software we use to extract information from the World Wide Web is similar. It allows us to query for particular information and returns the information in an orderly way so we can process it one link at a time. If all the information is available all the time, we cannot possibly process it all at the same time. This amount of information, without some interface, would confuse and overwhelm us.

So the Balance of One is a buffer, in many ways a protection, since an unprepared mind couldn't process this flood of information from all spacetime. It would overwhelm us psychologically and physically, causing aliments in both these realms.

But an awakened person is able to bypass the Balance of One filter intentionally, due to the fact that they have developed the needed psychological and physical circuitry necessary to process these new dimensions in a controlled and productive manner. Continuing the analogy, they do not need software to bring back information from the Time dimensions in an orderly manner; they can process multiple links instantly. They thereby acquire access to more of the Time Information Fields, and with this new insight, comes more responsibility.

Without the properly developed 'circuitry' within our consciousness and our biological structure, there is the possibility of psychological turmoil due to the necessity of holding two interlinked multidimensional realities, *this duality*, consciously at all times. While much of this interfacing into Time is done unconsciously for us, when it is done consciously, there is ownership of the process, and individuals can interfere with the automated unconscious activity of balance to their benefit or to their peril. The individual has to 'know thyself' and have a strong internal compass about where they choose to go with this new information. In 4D Time, I would expect the statement is, *Information is energy, knowing is power.* The true power is in knowing who you truly are.

One manner in which people have accessed more of the information about themselves in 4D Time is to be constrained in 4D Space, as a political prisoner, for example, or in a meditative state or retreat. Without much mobility in 4D Space, the option is to move more in 4D Time. But motion in and of itself is not automatically healthy. People can move into places that are quite dark in time, and suffer because of it. Others with the strength avoid these dark places and instead move to places inside of themselves that maybe they would not have accessed before, had they not been constrained. This exercise, done with intention and purpose over a period of time, typically years, allows one to master the 4D Time world inside. When a person masters this, they are truly masters of themselves, since they have truly come to know themselves through the challenges and opportunities of their circumstances.

The more into the One moment a person moves, the less the past and future influence their thoughts and actions. As the past and future influences lessen in their mental, emotional, and spiritual experiences, the hold that this duality of spacetime has on them begins to fade, and they can be described as being 'in this world, but not of this world.' Currently, this saying is based on a narrow definition of spacetime, that is, only three dimensions of space and one dimension of time. With an expanded concept of spacetime, this same statement can be viewed as one who is consciously bringing more 'time' into their everyday experience, balancing it with their conscious experience in space. Eventually, with no energetic attachment to experiences in 4D Space, or 4D Time, the individual becomes at One with the part of themselves outside of 4D Space and 4D Time, reaching what is commonly called en***light***enment.

12 Epilogue Multidimensional Time

Until you value yourself, you won't value your time. Until you value your time, you will not do anything with it.

M. Scott Peck

I hope that I have taken you on an interesting journey into multidimensional 4D Space and 4D Time. These new ideas about an expanded concept of time left me with more questions than I had before. I expect that these concepts will be no different for you, but what is important is that we continue to ask questions.

I remember my understanding about space and time distinctly being shaken a number of times due to ideas from the Theory of Relativity, to the relationship of the energies of time and space to spin angular momentum, to the speed of light as a ratio of space to time, to the theories of Zero-Point Energy, as well as the new concepts about human consciousness.

I spent years playing with these ideas on my own time. The simple understanding that time is energy and the concept that the speed of light is a ratio where space equals one divided by time and time equals one divided by space changes how you see all of spacetime. As 4D Space expands, 4D Time contracts and vice versa. That conservation of charge includes total electric and magnetic charge conservation in space and in time. The nice symmetry of electrodynamics when the physics of spacetime is expanded to eight dimensions. These insights opened my perception of the universe around me.

There could well be more dimensions, and likely are, but the amount of work necessary to understand the impact of just these extra four dimensions will certainly keep me busy for the rest of my days as a curious human.

The statement by Scott M Peck at the beginning of this epilogue correlates the link between time and consciousness in a clever way, but one that reflects accurately how we value ourselves and time.

In this journey in time, the concept of time as being *everywhere*, literally, in our universe and that we are tied to this *everywhere* helped me understand that I am truly a multidimensional spacetime being in a conscious universe, one that is evolving with every switch between states of 4D Space and states of 4D Time.

This expanded understanding of ourselves seems to link well with philosophical ideas about how we as humans are connected to the universe around us, but in the model presented in this book, it has a much more concrete and definable relationship.

There might be those who find the idea that our consciousness and emotions can be reduced to relations, or equations, of space and time a bit distasteful. But what is music if not just one complex mathematical relationship of time? The frequency of the notes and chords are one divided by time, and the beats are measured in periods of time. Music has the power to affect us deeply, and it is really just beautiful mathematical ratios constructed in time.

The question, then, is: if we are so connected, why can we not influence the universe more easily? As we increase the numbers of humans on planet Earth, this question is easy to answer. As a collective, our power to change our environment is clearly obvious. But the question of impact is also more personal. We know with enough intention and attention that we can change our conscious and unconscious selves. But if we are so connected, why can we not affect matter directly?

The message of the enlightened masters is that we do influence the universe, and can affect matter directly. Our acceptance of the fact that we consider ourselves separate, not only from each other, but from the universe, is said to be an illusion, obviously a very well-crafted illusion. But the message from spiritual masters has been that it takes great mental, emotional, and spiritual discipline and dedication to break through this illusion. The level of training and skill required to pilot a vehicle at 60 mph on a road is radically different than to pilot one at 17,500 mph in space.

If the vehicle is your consciousness, the level of understanding has to be vastly greater, since at a similar state to the 17,500 mph, every thought and feeling would generate vast amounts of energy. If you are not skilled enough in piloting your consciousness, it could become a catastrophic failure very quickly. In order to own this level of great creativity, you must own the potential for the opposite, the potentials for great failure. The only separation between the two seems to be your internal guidance system, that is, the part of you in time.

The question now is whether there are dormant structures in our 4D Space and 4D Time 'bodies and fields' that can be activated to allow much more interface into the creative energies of the universe. I believe so, and we as a human race are just beginning to glimpse what that means. We now have incredible knowledge and global contact right in the palm of our hands. It was not that long ago that access to limited amounts of knowledge required a real physical effort, as well as financial expense. Contacting people globally, especially in some regions of the Earth was challenging both in space and in time. Now the rate of access to incredible knowledge and to people all over the globe is expanding exponentially. With this new access, comes other issues and problems that we are currently dealing with, but the advantages outweigh the disadvantages, so we as humans clamor for more.

It is not difficult to 'feel' that humanity is at a critical junction. We have made tremendous leaps in technology, but our conscious awareness of the responsibility we hold for our technological creations seems to lag behind instead of us driving the technology for our clear benefit. Nowhere is this more obvious than in the detrimental effects we have allowed our technologies to make to our waters, our lands, our air, and to each other.

In the same way that technology has brought us this tremendous access to known information, I believe that we as humans have access to information that is available in time. I believe intuitive feelings, gut feelings, lucid dreams, near death experiences, and insights during meditation are precursors to our next stage of information access. We need to learn how to interpret this information, because it is formatted in a manner that is not based on linear thinking. The past and future seem to collapse into the present in ways that it would be beneficial to understand.

Einstein stated, '*We cannot solve our problems with the same thinking we used when we created them.*' Our world is rapidly challenging us as global citizens to see ourselves, and our responsibilities to each other, in new ways. We feel more connected to each other now than ever before, and technology has helped foster this. This new connected consciousness will help to bring about new solutions. Understanding how 'real' this connectedness is, beyond just technology, will help even more.

The opportunities that await us are tremendous, but only if we do not let fear lead us. When you consider that every thought and feeling can be as real as the buildings we live and work in, it means we need to take full responsibility of our own ‘emotional house,’ since it impacts all the Time Information Fields around us, including as global citizens.

13 References

Chapter 1

1. Scientific American, January 1986, "Applications of Optical Phase Conjugation," David M. Pepper
2. *About Time*, Paul Davis, Simon and Schuster 1995, p 204.
3. *The Holographic Universe*, Michael Talbot, HarperPerennial, 1991, p191-2.

Chapter 3

1. Everything for Nothing, Harold Putoff, p 1-3.
2. Quantum fluctuations of empty space: A new Rosetta Stone in physics? Dr. Harold Puthoff, Institute for Advanced Studies.

Chapter 6

1. *Nothing But Motion*, Dewey Larson, North Pacific Publishers, 1979, Chapters 2 and 6.
2. *Faster Than Light*, Nick Herbert, a Plume Book, 1989, p132.
3. *Nothing But Motion*, Dewey Larson, North Pacific Publishers, 1979, p148.
4. Beyond $E=mc^2$, Bernhard Haisch, Alfonso Rueda, H.E. Putoff, The Sciences, Vol. 34, no. 6, p 26-31, Nov/Dec 1994.
5. *Energy from the Vacuum*, Thomas E Bearden, Cheniere Press, 2008, p85.
6. *CYMATICS*, Hans Jenny, MACROmedia, 2001, p 237,127, and 47.

Chapter 7.

1. Everything for Nothing, Harold Putoff, p 1-3.
2. *Energy from the Vacuum*, Thomas E Bearden, Cheniere Press, 2008, p88.

3. Ibid, p128.
4. Ibid, p129.
5. Ibid, p131.
6. Ibid, p141.
7. Introduction to Zero Point Energy, Calphysics Institute, http://www.calphysics.org/zpe.html
8. Measurability of vacuum fluctuations and dark energy, Christian Beck, Michael C. Mackey, Dec 11, 2006, http://arxiv.org/pdf/astro-ph/0605418.pdf

Chapter 8

1. Detection of a Dipole in the Handedness of Spiral Galaxies with Redshifts z ~0.04, (Michael J. Longo), *Phys. Lett.* **B 699**, 224-229 (2011).
2. www.letstalkaboutscience.com.
3. *Energy from the Vacuum*, Thomas E Bearden, Cheniere Press, 2008, p136.
4. *Energy from the Vacuum*, Thomas E Bearden, Cheniere Press, 2008, p141.
5. Ibid.

Chapter 9

1. *Holographic Universe*, Michael Talbot, HarperPerennial, 1991, p11-14.
2. Karen K. DeValois, Russell L. DeValois, and W. W. Yund. "Responses of Striate Cortex Cells to Grating and Checkerboard Patterns," *Joumal of Physiology*, vol 291, 483-505, 1979.
3. *Holographic Universe*, Michael Talbot, HarperPerennial, 1991, p90-101.

Chapter 10

1. *Entangled Minds*, Dean Radin, Paraview Pocket Books, 2006, p268.
2. Brain States, pysc101, wiki
3. The Energetic Heart. Bioelectrical Interactions Within and Between People, Rollin McCraty, Institute of HeartMath, p1.
4. Ibid, p3.

5. Ibid, p2.
6. Ibid, p11.
7. Coherence, Bridging Personal, Social, and Global Health, Rolin McCraty PhD, Doc Childre, p10.
8. The Energetic Heart. Bioelectrical Interactions Within and Between People, Rollin McCraty, Institute of HeartMath, p4.
9. Coherence, Bridging Personal, Social, and Global Health, Rolin McCraty and Doc Childre. p21.
10. *The Body Electric*, Robert Becker, William Morrow, 1985, p245.
11. Ibid, p244.
12. Ibid, p246.
13. Ibid, p248.
14. Ibid, p247-9.
15. *Morphic Resonance*, Rupert Sheldrake, Park Street Press, 2009, p52.
16. Ibid, p25.
17. Ibid, p8.
18. Ibid, p61.
19. Ibid, p29.
20. Ibid, pxxv.

Chapter 11

1. *Nothing But Motion*, Dewey Larson, North Pacific Publishers, 1979, p 33-35.

14 Part 2: APPENDIX

This section is included for those who wish to dig deeper into the technical details of what the extra dimensions of time and space might mean for the current theories. I took the ideas of Dewey Larson regarding space and time and applied them to electrodynamics. These derivations produced some very interesting results and gave me a whole new insight into spacetime from an electric and magnetic point of view. These papers are available at my website www.multidimensionaltime.com

These sections in the Appendix will hopefully inspire those readers who are interested in the ideas presented in this book and want to explore further. These papers show work that I have done based on my own curiosity, and my hope is that it will inspire others to do their own investigations. These papers reflect my own investigations and are not intended to reflect the viewpoints of the Reciprocal System Theory.

15 A 1 - Concepts of Three-Dimensional Vector Time in Electrodynamics

Robert Kersten
www.multidimensionaltime.com

November 21, 2014

Abstract.

Spacetime in physics is represented as a four-dimensional vector, with time as a scalar and space as a three-dimensional vector. But is there more dimensionality to spacetime, for instance, three-dimensional vector time and scalar space? Having a location in 3D vector time is easy enough to imagine, and so is moving in time. When the mathematics of moving in 3D vector time are laid out, it forces some new insights into the relationship and symmetry between Space and Time. The idea of Time, Space, and velocity as a coupler between the two is postulated in the Reciprocal System Theory. In this paper, instead of postulates, a mathematical approach is laid out using quaternions to derive properties of 3D vector time, 3D vector space, and how velocity couples the two. This mathematical approach is then applied to electrodynamics, resulting in four additional equations in 3D vector time and scalar space. This new model of three dimensions of vector space and time added to three dimensions of vector time and scalar space gives a total of eight equations for electrodynamics.

15.1 Introduction

The concept of location in 3D vector space is very familiar to us. So is the concept of moving in 3D vector space. But to move in 3D vector space requires time. Without at least one dimension of time, it is not possible to move in space, since movement is change in space divided by change in time. A change in space divided by a change in space is not movement.

The concept of 3D vector time can be imagined, and we can even imagine moving from one location in time to another, just like in three-dimensional vector space. But when 3D vector time is put into mathematics, the idea of moving in time becomes an issue. A change in time divided by a change in time is not movement. Just like three-dimensional vector space, to move in time requires a change in time and a change in space, specifically a change in time divided by a change in space, or the reciprocal of velocity in 3D vector space.

The speed of light is a ratio of space to time, increasing time has the same effect as decreasing space and vice versa. This reciprocal relationship between space and time will be used to couple space and time in a new way, where time is a three-dimensional vector and space is a scalar.

What does 3D vector time mean? We intuitively understand one dimension of time, but what about two dimensions of time [t^2], or even three dimensions [t^3]. Velocity and reciprocal velocity will be used to gain understandings of these new dimensions in vector time. Dewey Larson, in his Reciprocal System Theory, has raised some very interesting ideas on the concept of three-dimensional vector time and the coupling of time and space via velocity[1]. In the Reciprocal System Theory, the lack of any mathematical derivation in the Reciprocal System can make it hard to follow how his concepts are derived. This paper proposes a mathematical approach to the idea of three-dimensional vector time with velocity as a coupler between space and time and look at what impact this has on electrodynamics.

The form used in this paper are quaternions, the same form that Minkowski used to formulate spacetime with four dimensions, a three-dimensional vector for space and a scalar form of time. Quaternions were developed by W.R. Hamilton and were used by James Maxwell when he formulated his original Electrodynamic equations. In years after his work, Oliver Heaviside found that reducing a four-dimensional quaternion to a three-dimensional vector simplified the equations. The rationale for eliminating the scalar component is that the scalar portion had no physical analog, so could be ignored in the interest of simplicity. In this paper, as well as others to follow, the scalar portion has significance. The quaternion form that will be used in this paper is the complex quaternion, or bi-quaternion [2]. In this notation, all the scalars and vectors are complex. The bi-quaternion form of the quaternion is not explicitly required for the derivations in this paper, but I want to start with this format to allow maximum flexibility for future developments.

The Minkowski spacetime quaternion is significant because it couples space and time together into a 4D spacetime vector. It changed the way we looked at the universe. But is there still more to be gained from a reciprocal of this known spacetime quaternion, a new quaternion where the vector is a three-dimensional time and the scalar variable is space? These additional four dimensions of spacetime and how it applies to electrodynamics is the reason for this paper.

To start off, for clarity, dimensions of space and time will be treated separately in the first two sections and then combined after that.

Note: In this paper, inverse and reciprocal mean the same thing.

15.2 Three-Dimension Vector Space [3DS]

In 3D vector Space, or 3DS, the dimensions of length [L], area [L^2] and [L^3] are very familiar to us. These variables define static space. As stated previously, to move from one location in space to another, it cannot be done in space since a change in space relative to space is no change in location, and therefore no motion has been achieved.

$$d/dx\ \mathbf{x} = 1, \text{ or } \nabla \mathbf{x} = 3 \quad \text{where } \mathbf{x} = x_1\mathbf{x_1} + x_2\mathbf{x_2} + x_3\mathbf{x_3} \quad (1)$$

To reiterate, in order to move in space, a change in time is required, in other words dx/dt or a velocity. This exercise will be done in detail to elucidate concepts in 3DS, which will be useful when ideas in three-dimensional time are explored in the next section. Velocity in one dimension is represented as

$$d/dt\ \mathbf{x} = d\mathbf{x}/dt \qquad (2)$$

and in three dimensions

$$d/dt\ \mathbf{x} = \mathbf{v} \quad \text{where } \mathbf{v} = v_1\, \mathbf{x_1} + v_2\, \mathbf{x_2} + v_3\, \mathbf{x_3} \qquad (3)$$

Here, any movement is expressed in units of space/time [s/t] or meters/second. It is important to define units in terms of space and time since it will help to understand relationships later.

To get acceleration, again, a change in time is necessary.

$$d/dt\ \mathbf{v} = d\mathbf{v}/dt \text{ or acceleration} \qquad (4)$$

where $\mathbf{a} = a_1\, \mathbf{x_1} + a_2\, \mathbf{x_2} + a_3\, \mathbf{x_3}$

So to move from one location to another, either at constant velocity or at an ever-increasing velocity requires a change in time. Units for acceleration are space/time^2 [s/ t^2] or meters/second^2.

15.3 Space in Quaternion Form

The spacetime quaternion for a location in space is

$$\mathcal{X} = (x, \boldsymbol{i}\cdot\boldsymbol{x}) \quad \text{where x is } x_0 + i\, x_0 \text{ and } \boldsymbol{i}\cdot\boldsymbol{x} \text{ is } \boldsymbol{i}\cdot(\boldsymbol{x} + i\boldsymbol{x}) \quad (5)$$

where x is the complex scalar location and $\boldsymbol{i}\cdot\boldsymbol{x}$ the complex vector representing the location. For a three-dimensional position vector, the complex scalar is equal to zero, and the complex coefficients of the vector terms are zero. For derivations with quaternions, the operator Nabla ∇ is used. For this paper, when Nabla is used in reference to 3DS, I have put a subscript s on the operator, ∇_s . The subscript s is used to denote that the derivation in vector form is with respect to space. Nabla for space is

$$\nabla_s = (i/c\ \partial/\partial t + \boldsymbol{i}\cdot\nabla_s) \quad \text{where } \nabla_s = (\partial/\partial x_1 + \partial/\partial x_2 + \partial/\partial x_3) \quad (6)$$

In bi-quaternion form, a total derivative d/dt [3]

$$d/dt = (\partial/\partial t,\ -\boldsymbol{v}\,\nabla_s) \quad (7)$$

where $-\boldsymbol{v}\,\nabla_s$ is the vector derivation with respect to time, as is shown below. [2]

$$-\boldsymbol{v}\,\nabla_s = -\,dx/dt\ \partial/\partial x + dy/dt\ \partial/\partial y + dz/dt\ \partial/\partial z = d/dt \quad (8)$$

So, to repeat the exercise of moving in space, the vector derivative of the position quaternion with respect to time is

$$d/dt\ \mathcal{X} = (\partial/\partial t,\ -\boldsymbol{v}\,\nabla_s)(x, \boldsymbol{i}\cdot\boldsymbol{x}) \quad (9)$$

$$V_S = d\,X/dt = (\partial/\partial t, -\,\mathbf{v}\,\nabla_s X) \text{ or } V_S = (v, -\,\mathbf{v}) \quad (10)$$

Where the quaternion velocity V_S *is* in units of space/time or s/t. In free space,

$$V_S = (c, -\,\boldsymbol{i}\cdot\mathbf{v}) \quad \text{where c is now the speed of light} \quad (11)$$

To finish the exercise, we compute bi-quaternion acceleration.

$$A_S = (a, \boldsymbol{i}\cdot\mathbf{a}) \quad (12)$$

15.4 Three-Dimensional Vector Time [3DT]

The concept of three-dimensional vector Time, or 3DT, is a bit foreign since we are so accustomed to thinking of time in terms of a scalar variable. It can move forward and backward, but universally, we think of time as a scalar that moves forward relentlessly.

To mathematically represent 3D vector time is easy. The dimension of time in 1D is [t], in 2D time [t^2] and in 3D [t^3] . We have no understanding of what these units mean. Let's start by representing a location in 3D vector time. The time vector is represented as follows:

$$\boldsymbol{t} = t_1\,\boldsymbol{t_1} + t_2\boldsymbol{t_2} + t_3\boldsymbol{t_3} \quad (13)$$

Let's first go through the same exercise as before, where we start at a location in time and try to move in time. If we apply a delta t to our location in time, we have the same problem as in space.

$d/dt\ \mathbf{t} = 1$, or $\nabla_t\ \mathbf{t} = 3$ where $\nabla_t = (\partial/\partial t_1 + \partial/\partial t_2 + \partial/\partial t_3)$ (14)

There is no change in the location in time, no motion. In order to move in time, a change in space is required.

$d/dx\ \mathbf{t} = d\mathbf{t}/dx$ or reciprocal velocity (15)

and in three dimensions of vector space

$d/ds\ \mathbf{t} = d\mathbf{t}/ds = 1/\mathbf{v}$ or reciprocal velocity in 3DT. (16)

So velocity in time, in vector form, is represented as

$Velocity\ (in\ time) = 1/v_{t1}\ \mathbf{t_1} + 1/v_{t2}\ \mathbf{t_2} + 1/v_{t3}\ \mathbf{t_3}$ (17)

The units would be time/space or t/s [seconds/meter]. So in 3DT, movement is done in reciprocal velocity relative to our common experience of velocity in 3DS. To finish up our derivation of a change in 3DT, we now determine how to move faster between locations in 3DT.

To have a change in velocity, or acceleration, another change in space is required.

$d/ds\ \mathbf{v} = d\mathbf{v}/ds = \mathbf{a}$ (18)

where $\mathbf{a} = 1/a_{t1}\ \mathbf{t_1} + 1/a_{t2}\ \mathbf{t_2} + 1/a_{t3}\ \mathbf{t_3}$

The units for acceleration in 3D vector time are t/s^2[seconds/meter^2]. To complete the exercise, we complete quaternions for location, velocity, and acceleration in 3DT.

15.5 Time in Quaternion form

So bi-quaternion form of a location in 3DT is

$$\mathcal{T} = (f(t),\ \boldsymbol{i \cdot t}) \tag{19}$$

where t is $t_0 + i\ t_0$ and $\boldsymbol{i \cdot t}$ *is* $\boldsymbol{i \cdot (t + it)}$

where *f(t)* is a scalar function of time and **t** the vector to the location in time.

To move in 3DT, we need a new definition of the total derivative for 3DT with respect to space.

$$d/ds = (\partial/\partial s,\ -\ \boldsymbol{(1/v)}\ \nabla_t) \tag{20}$$

which is the reciprocal of the total derivative d/dt shown in equation seven.

$$-(1/\boldsymbol{v})\ \nabla_t\ =\ d\ t_1/ds\ \partial/\partial t_1 +\ d\ t_2/ds\ \partial/\partial\ t_2\ + d\ t_3/ds\ \partial/\partial\ t_3$$
$$= d/ds \tag{21}$$

So, bi-quaternion Velocity in 3DT is

$$d/ds\ \mathcal{T} = (\partial/\partial s,\ -(1/\boldsymbol{v})\ \nabla_t)(\ f(t),\ \boldsymbol{i \cdot t}) \tag{22}$$

$$d/ds\ \mathcal{T} = \mathbb{V}_t = ((\partial f(t)/\partial s,\ -(1/\boldsymbol{v})\ \ d/ds(\boldsymbol{i \cdot t})$$

$$\mathbb{V}_{\mathbf{t}} = (1/v,\ -1/\ \boldsymbol{i \cdot v}) \tag{23}$$

where v is speed, likely c, and **v** is velocity

where $\mathbb{V}_t$ denotes velocity in 3DT. The bi-quaternion acceleration $\mathbb{A}_t$ in 3DT is

$$\mathbb{A}_t = (1/a,\ 1/\ \boldsymbol{i \cdot a}) \tag{24}$$

15.6 How to combine it all.

So far, space and time have been evaluated separately, to clarify the concepts involved in three-dimensional time. Physics has space and time coupled and as shown in the previous sections. To move in spacetime now requires either dx/dt or dt/dx. The two quaternions, $\mathbb{X}$ and $\mathbb{T}$, need to be coupled. These two coupled four-dimensional representations of spacetime will result in a total of eight spacetime dimensions.

In the Minkowski representation of spacetime in physics, there is a scalar time and a vector space.

$$\mathbb{X}(t) = (c\, f(t), \boldsymbol{i}\cdot\boldsymbol{x}) \text{ where } f(t) \text{ is a scalar time field.} \quad (25)$$

I will call this 4D spacetime representation 4D Space.

There are mathematical operators that allow us to go from scalar to vector and vector to scalar. The mathematical operation divergence, represented here as $\nabla_s \cdot$ and $\nabla_t \cdot$, will give us a scalar field value from a space vector in the former and a time vector in the latter. The result of this divergence operator will determine if the scalar field is a source or a sink. If scalar time is a source in three-dimensional space, then, in order to conserve energy in these eight dimensions, scalar space would have to be a sink in three-dimensional time, or vice versa. The gradient operators, ∇_s and ∇_t, give a vector from a scalar. So if we apply the time gradient operator to the time scalar field value of equation 25, where t is a function of scalar time, a vector form of **t** is the result.

$$\nabla_t (f(t)/c) = \boldsymbol{i}\cdot\boldsymbol{t} \quad (26)$$

where ∇_t is the gradient with respect to time

If the space divergence operator is applied to the space vector in equation 25, scalar field of space is calculated. The results for scalar field space is shown next, where **v** equals the speed of light

$$\nabla_s \cdot \boldsymbol{i}\cdot\boldsymbol{x} = f(s) \tag{27}$$

where f(s) is a scalar space field

Combing the results, the quaternion representation of spacetime for 4D Time is

$$\mathbb{T} = (f(s)/c, \mathbf{i}\cdot\mathbf{t}) \tag{28}$$

This will be labeled 4D Time, for the quaternion in the vector time portion of spacetime. The two quaternions are coupled via divergence and gradient, as shown in the next equation.

$$\mathbb{T}(s,\mathbf{t}) = ((\nabla_s \cdot \ \mathbb{X}(t,\mathbf{s})\ )), \boldsymbol{i}\cdot(\nabla_t\, \mathbb{X}(t,\mathbf{s})\ )) = (f(s)/c\ , \boldsymbol{i}\cdot\boldsymbol{t}) \tag{29}$$

and

$$\mathbb{X}(t,\mathbf{s}) = ((\nabla_t \cdot \ \mathbb{T}(s,t))\ , \boldsymbol{i}\cdot(\nabla_s\, \mathbb{T}(s,t)) = (c\, f(t)\ , \boldsymbol{i}\cdot\boldsymbol{x}\ ) \tag{30}$$

From there, we can take the total derivative with respect to space to get velocity and acceleration in time.

15.7 Implications for Speed of Light

In relativity, the interval is invariant, so this property will be used.

$$t' = (t^2 - x^2) \tag{31}$$

where in 4D Space

$$x = v\, t \tag{32}$$

so putting the latter equation into the previous, the interval is

$t' = t(1 - v^2) = t(1 - v^2/\, c^2) = t$ where $v<c$ to keep t from going negative. (33)

$$d\, \mathbb{X}(t,\mathbf{s})/dt = \gamma_s\, (-ic,\, \boldsymbol{i \cdot v}) = \textit{max of } c \tag{34}$$
where $\gamma_s = (1 - v^2/\, c^2)^{1/2}$

For relativistic velocity in 4D Time, v is dt/dx, so x in equation 32 changes to

$$x = t/v \tag{35}$$

$t' = t(1 - 1/v^2) = t(1 - c^2/\, v^2) = t$ where $v>c$ to keep t from going negative. (36)

so the new γ_t , is

$$\gamma_t = (1 - c^2/\, v^2)^{1/2} \tag{37}$$

$$d\, \mathbb{T}(s,\mathbf{t})/ds = \gamma_t\, (-i/c,\, \boldsymbol{i \cdot 1/v}) = \textit{min of } c \tag{38}$$

The result is that in 4D Time, velocity is always superluminal. For 4D Time, the Lorentz length is negative for $v<c$ and expands for $v>c$. The time contracts and length expands for $v>c$, the inverse of familiar length contraction and time dilation of 4D Space.

15.8 Velocity as a coupling

So far, there are two spacetime quaternions. The first one, $\mathbb{X}(t,\mathbf{s})$, couples scalar time field and vector space field. A scalar field is defined as a field for which there is a single value for each point in the field. The vector field has a magnitude and direction for each point in the field. The second one, $\mathbb{T}(s,\mathbf{t})$, couples scalar space field and vector time field. Locations in 4D Space [$\mathbb{X}(t,\mathbf{s})$] are in units of space, or s. In 4D Time [$\mathbb{T}(s,\mathbf{t})$], locations in time are in seconds, which will be represented as t for time. The reason for this unit representation is so velocity can be represented for what it is, space/time, or s/t. To convert a location in 4D Space using time, velocity is needed to keep the units in space, as in, *ct*. In 4D Time, $\mathbb{X}(t,\mathbf{s})$ is multiplied by the 1/c to keep the units in time, so, *x/c.*

(Location in) time * (c)space/time = (Location in) space → t * s/t = s (39)

(Location in) space * (1/v) time/space = (Location in) time → s * t/s = t (40)

Previously, it was just a location in spacetime. Now there is movement in spacetime, or velocity. To change from velocity in 4D Space to velocity in 4D Time, the reciprocal velocity squared is required.

velocity (in space) x $1/v^2$ = reciprocal velocity(in time) (41)

Looking at the coupling in units, it gives a better insight

$s/t * t^2/s^2 = t/s$ (42)

where t/s is reciprocal velocity in 4D Time

For accelerations, reciprocal velocity cubed is required. If we take equation 41 and take the time derivative of velocity, we get acceleration [units s/t^2]. The space derivative of t/s, that is, d/ds t/s is equal to t/s^2 and the coupling between the two of reciprocal velocity cubed. The equation becomes,

$$s/t^2 * t^3/s^3 = - t/s^2 \qquad (43)$$

The most important point of this section that is scalar time and vector space [4D Space] and scalar space and vector time [4D Time] are coupled via velocity and reciprocal velocity to powers of 1, 2, and 3. This coupling is crucial for the next section and for future understanding.

Dewey Larson postulated that t^3/s^3, or $1/c^3$ is mass. The term s/t^2 is the familiar acceleration in vector space. The idea of not treating mass as a physical property was raised in a number of papers on mass, inertia, and ZPF. In one particular paper, two particular statements are very important. The first statement is: "*In our formulation, the m in Newton's second law of motion, F=ma, becomes nothing more than a coupling constant between acceleration and an external electromagnetic force.*" The second statement deals with another famous equation, $E=mc^2$ and states "*Mass is energy.*" [5] With the middle term of equation 43, t^3/s^3, equivalent to mass and used as a coupling term, this implies that the term t/s^2 is Force, in terms of vector time. With force equal to t/s^2, then Energy is Force times distance, or

$$\text{time/space}^2 \times \text{space [distance]} = \text{time/space or } 1/c \textit{ in units of } t/s \qquad (44)$$

Energy, then, is equivalent to t/s, or inverse velocity 1/c. Using the equation $E=mc^2$ also confirms the same conclusion that energy is equal to 1/c. This means if the total derivative, *d/ds,* of any location in vector time, it is equivalent to vector Energy. The bi-quaternion representation for Energy is

$d/ds\ \mathcal{T} = (e\ ,\ \mathbf{E})\quad \mathbf{E} = E_{t1}\ \mathbf{t_1} + E_{t2}\ \mathbf{t_2} + E_{t3}\ \mathbf{t_3}$ *in units of* t/s (45)

A scalar energy is familiar in 4D Space, but a vector energy is a new concept. The scalar energy is the portion of motion in three-dimensional time that shows up in three-dimensional space. The divergence operator compresses this three-dimensional motion into a scalar form in three-dimensional space.

If compounded motion 4D Time is vector Force, that is the total derivative taken twice - d^2/ds^2, and compound motion in 4D Space is vector acceleration [d^2/dt^2], then the bi-quaternion representation of force is

$d^2/ds^2\ \mathcal{T} = (f\ ,\ \mathbf{F})\quad \mathbf{F} = F_{t1}\ \mathbf{t_1} + F_{t2}\ \mathbf{t_2} + F_{t3}\ \mathbf{t_3}$ *in units of* t/s^2 (46)

In this equation, it is vector force that is familiar in vector space and scalar force that is not. In this case, the scalar force should always equal zero, since a uniformly diverging or converging force field sums to zero.

15.9 Dual Symmetry in Maxwell equations

When Maxwell's equations are evaluated for free space, the symmetry between electrical and magnetic fields is very apparent, as shown in the four equations below [CGS units].

$$\nabla \cdot E = 0 \quad (47)$$

$$\nabla \cdot B = 0 \quad (48)$$

$$\nabla \times E + 1/c\ \partial B/\partial t = 0 \quad (49)$$

$$\nabla \times B - 1/c\ \partial E/\partial t = 0 \quad (50)$$

It can be seen that E equals B and B equals -E. There is velocity coupling the two fields together.

This symmetry between E and B fields and the derivative with respect to space for one property and with respect to time for the other property has a similar symmetry between the Mikowski four-dimensional space quaternion and its reciprocal, the four-dimensional time quaternion.

For Maxwell's equations not in free space, charges need to be considered and the symmetry between E and B is broken. The most obvious asymmetry is that static electric fields are derived from a source, a charge density, and magnetic fields are not. In the dynamic equations, to calculate a dynamic magnetic field, an electric current is required, along with a time-changing electric field, where, in magnetic fields, only a time-changing electric field is required.

For clarification, ∇ is now denoted as ∇_s where the subscript refers to 3D vector space, to make a distinction from ∇_t which refers to 3D vector time. This subscript is true for the electric and magnetic fields as well. For 3D Space, Maxwell's equations in CGS units are

$$\nabla_s \cdot E_s = 4\pi\rho_e \tag{51}$$

$$\nabla_s \cdot B_s = 0 \tag{52}$$

$$\nabla_s \times E_s + 1/c\ \partial B_s/\partial t = 0 \tag{53}$$

$$\nabla_s \times B_s = 1/c\ (4\pi J_e + \partial E_s/\partial t) \tag{54}$$

In Dual Symmetry, where magnetic monopoles are assumed to exist and have not been measured yet, the Dual Symmetry expands to

$$E_s \rightarrow B_s \quad \text{and} \quad B_s \rightarrow -E_s \tag{55}$$

$J_{es} \rightarrow J_{ms}$ and $J_{ms} \rightarrow -J_{es}$ (56)

$\rho_{es} \rightarrow \rho_{ms}$ and $\rho_{ms} \rightarrow -\rho_{es}$ (57)

Using these symmetries, Maxwell's equations become

$$\nabla_s \cdot E_s = 4\pi \rho_{es} \quad \text{in V/m}^2 \quad (58)$$

$$\nabla_s \cdot B_s = 4\pi \rho_{ms} \quad \text{in V/m}^2 \quad (59)$$

$$\nabla_s \times E_s + 1/c\ \partial B_s/\partial t = 1/c\ J_{ms} \quad \text{in V/m}^2 \quad (60)$$

$$\nabla_s \times B_s - 1/c^2\ \partial E_s/\partial t = 1/c\ (4\pi J_e) \quad \text{in V/m}^2 \quad (61)$$

But no magnetic monopoles have been detected in three-dimensional space and scalar time.

15.10 Electrodynamic equations in 3D Vector Time

Using the remarkable symmetry to the coupling of 4D Space to 4D Time using velocity and reciprocal velocity, this approach will be used to evaluate electrodynamics in three-dimensional time and scalar space. As a reminder, area in vector time is seconds^2, volume is seconds^3 and movement in time requires a derivative with respect to space. Potentials are common to both vector space and vector time.

In this section, the electrodynamic equations will be evaluated in three-dimensional vector time and scalar space. When the transformation from vector space and scalar time to vector time and scalar space is done, in both units of CGS and SI, it becomes clear that the electric field is derived from electric monopoles in vector space, but in vector time, the electric field only has a flux component, meaning that electric monopoles are not part of vector time. For the magnetic field, the derivations show the magnetic field as a charge density in vector time, so magnetic fields are derived from magnetic monopoles. This fact becomes much clearer when the electrodynamic symmetry is analyzed with SI units.

This symmetry is first investigated in CGS units, since the speed of light is explicit in the equations. In SI units, the speed of light is buried in the permittivity and permeability terms. In addition, the new units in CGS units are less complicated, and so this is more easily elucidated in these units.

The electromagnetic fields of 4D Space are correlated to 4D Time in the following ways:

$$E_{time} = c\, E_{space} \quad \text{and} \quad B_{time} = c\, B_{space} \tag{62}$$

Also, for the derivatives,

$$\nabla_s = \nabla_t / c \tag{63}$$

and

$$\partial/\partial t = \partial s/\partial t\ \partial/\partial s = c\ \partial/\partial s \tag{64}$$

So the electric and magnetic fields, which have units of Volts/meter in vector space, the units now become Volts/second in vector time.

15.11 Electrodynamic equations in 3D Vector Time with CGS units

In vector time, there are no electric monopoles, or charges, so electrical charge density $\rho_{et} = 0$ and $J_{et} = 0$. Also, the electric current density $J_{et} = 0$. But the magnetic charge density $\rho_{mt} \neq 0$ and the magnetic current density $J_{mt} \neq 0$. Therefore,

$$\nabla_t \cdot E_t = 0 \qquad \text{units are } V/s^2 \quad (65)$$

$$\nabla_t \cdot B_t = 4\pi\ c\ \rho_{mt} \qquad \text{units are } V/s^2 \quad (66)$$

$$\nabla_t \times E_t = 4\pi\ J_{mt} + c\partial B_t /\partial s \qquad \text{units are } V/s^2 \quad (67)$$

$$\nabla_t \times E_t = c^2\ \partial\ B_t /\partial s \qquad \text{units are } V/s^2 \quad (68)$$

15.12 Maxwell's equation in 3D Vector Time with SI units

For Maxwell's equations in SI units, the units of the electric field are now V/m and magnetic flux is V*s/ m^2. Permittivity (ε_s) is in units of A*s/V*m and permeability(μ_s) is in units of V*s/A*m, where and ε_s times μ_s equals $1/c^2$.

$$\nabla_s \cdot E_s = - \rho_{es}/\varepsilon_s \qquad \text{units of } V/m^2 \quad (69)$$

$$\nabla_s \cdot B_s = 0 \qquad \text{units of } Wb/m^3 \text{ or } V{*}s/\ m^3 \quad (70)$$

$$\nabla_s \times E_s + \partial B/\partial t = 0 \qquad \text{units of } V/m^2 \quad (71)$$

$$\nabla_s \times B_s = \mu_s\ J_{es} + 1/c^2\ \partial E/\partial t \qquad \text{units of } Wb/m^3 \quad (72)$$

Using the new degrees of freedom, the proposed form of Maxwell's equations for 4D Time and 4D Space are the following. Equations relating to the electric field charge density and equations relating to magnetic field [B] have flux density.

For SI units, the following relations between vector space and vector time are important. This relationship is the only way the units work out, and what the units suggest is very important.

$$E_t = c^2 E_s \quad \text{units of Volt*meter/second}^2 \tag{73}$$

In vector time [4D Time], the electric field in time is related to the electric field in space by the relationship $E_t = c^2 E_s$ which converts V/m to $V*m/s^2$. This changes the electric field from one based on charge density to an electric flux.

$$B_t = c^2 B_s \quad \text{units of Volts/second} \tag{74}$$

In 3D vector time, the magnetic field in time has to be related to the magnetic field in space by the relationship $B_t = c^2 B_s$ which converts $V*s/m^2$ to V/s. This changes magnetic flux to a magnetic field based on charge density.

For permittivity and permeability

$$\varepsilon_t = c^2 * \varepsilon_s = 1/\mu_s \quad \text{units of A*m/V*s} \tag{75}$$

$$\mu_t = c^2 * \mu_t = 1/\varepsilon_s \quad \text{units of V*m/A*s} \tag{76}$$

It is interesting to note that permittivity in 3D time becomes the inverse of permeability in 3D space. This is consistent with the electric field being a flux in 3D time. The same is true for permeability in time, which is the inverse of permittivity in 3D space.

$$\varepsilon_t * \mu_t = c^2 \tag{77}$$

It is important to remember that since these next four equations describe electrodynamics in 3D vector time, the basis units is per second instead of per meter as in 4D

Space since volume is measured in seconds instead of meters. Any movement in the scalar direction is time in 3D vector space and space in 3D vector time. The curl is taken with respect to time, and so the space derivate is taken for the variable in the scalar dimension. Since magnetic monopoles are in 3D vector time, the units for the magnetic field are V/s^2, that is charge density, and the units for the electric field are in terms of flux, so the units are $V*m/s^2$. In 3D vector time, Maxwell's equations become,

$$\nabla_t \cdot E_t = 0 \qquad \text{units of } V*m/s^3 \quad (78)$$

$$\nabla_t \cdot B_t = - 1/c^2 \mu_t \rho_{mt} \qquad \text{units are } V/s^2 \quad (79)$$

$$\nabla_t \times E_t - c^2 \partial B_t / \partial s = c J_{mt} / \varepsilon_t \qquad \text{units are } V*m/s^3 \quad (80)$$

$$\nabla_t \times B_t + \partial E_t / \partial s = 0 \qquad \text{units of } V/s^2 \quad (81)$$

When magnetic charge density, or magnetic monopoles, along with an electrical flux, are introduced in 3D vector time, a symmetry is created with the electrical charge density and magnetic flux in 3D vector space. In 3D vector space, ε_s is the electrical charge term, since electric fields have a source in charges and u_s is the magnetic flux. In 3D vector time, Electric fields are not derived from monopoles' charges and so E fields are treated as flux, so ε_t is the electrical flux term and u_t the magnetic charge term.

Electric monopoles are measurable in 4D Space. But magnetic monopoles are not, just like three-dimensional vector time is not yet measurable. From these derivations, once one is detected, the other will likely also be detectable.

15.13 Summary

These eight electrodynamic equations give the full symmetry between space and time, impacting how electric and magnetic fields relate to these extra dimensions of spacetime. These additional degrees of freedom, as radical as they might currently appear, allow for magnetic monopoles from a four-dimensional field of 3D vector time and scalar space, where motion is always faster than light.

The basis for magnetism in 3D vector space and 1D scalar time is electric charges. Magnetism in the form of magnetic dipoles are generated by either rotating charges, that is charges rotating in scalar time, or electric fields changed in scalar time. Without scalar time, there can be no magnetism. So 3D vector space can be described as electric space based on electric monopoles.

In 3D vector time and 1D scalar space, electricity in the form of electric dipoles are only generated by magnetic charges that move in scalar space. Without scalar space, in 3D vector time, there can be no electricity, so 3D vector time can be described as magnetic time based on magnetic monopoles.

Movement in time manifests as energy, and acceleration in time manifests as a force. The familiar scalar time field is 3D vector time compressed into 3D scalar time. This scalar field must have extremely high uniformity, since scalar time is used as a one-dimensional scalar variable without affecting the physics. It appears we have much more to learn about time. 3D vector time is easier to imagine, but movement in vector time requires scalar space, resulting in a reciprocal velocity from the perspective of 3D vector space. Just like the scalar time field manifests as a uniform scalar field, that is, as a uniform density in 3D space, so it could be imagined that scalar space field is a uniform density in three-dimensional vector time.

With so many extra degrees of freedom, it helps to keep an open mind and explore what the implication can be. It is clear that Physics does constrain time to one dimension. Representing time as multidimensional does not change the current physics, yet. But the extra four equations of electrodynamics that come from time in three dimensions and space as a scalar variable can open a whole new way of looking at the interactions of space and time.

References

1. *Nothing But Motion*, Dewey Larson, North Pacific Publishers, 1979. ISBN 0-913138-07-x. Information can also be accessed at www.reciprocalsystem.com/dbl/.
2. Application of Bi-Quaternion in Physics, Andre Waser, www.andre-waser.ch, June 05, 2007.
3. Electromagnetic Duality, Charges, Monopoles, Topology. Juan A. Mignaco, Brazilian Journal of Physics, vol 31, no 2, June 2001.
4. Inertia as a zero-point-field Lorentz force, Bernard Haisch, Alfonso Rueda and H.E. Putoff, Physics Letters A, Vol. 49, No 2, pp 678-694, 1993.
5. Beyond $E=mc^2$, Bernhard Haisch, Alfonso Rueda, H.E. Putoff, The Sciences, Vol. 34, no. 6, p 26-31, Nov/Dec 1994.

Version

1. Nov 21, 2014. Original.
2. Dec 29, 2014. Added in reference 5 and made reference to it in section 1.6.
3. Jan 02, 2015. Corrected typo in c term for equations 79 and 80
4. Jan 17, 2015. Add in vector in 3D time to differentiate it from 3D scalar time field.
5. Nov 30, 2015. In conclusion, add description of electric space and magnetic time.

16 A 2 - Electrodynamic Potentials in Three-Dimensional Vector Time

Robert Kersten
www.multidimensionaltime.com

January 02, 2015

Abstract.

Spacetime in physics is represented as a four-dimensional vector, with time as a scalar and space as a three-dimensional vector. But as shown in the paper 'Concepts of Three-Dimensional Time in Electrodynamics,' there is a time domain where time is a three-dimensional vector time and space is scalar, resulting in a total of eight equations for electrodynamics. In this paper, the well-known derivation of Maxwell's equations in terms of scalar and vector potentials is shown, and then the same approach is used to derive scalar and vector potentials for three-dimensional vector time and scalar space. The exercise of having all eight equations in terms of scalar and vector potential shows that the scalar wave equation is exactly the same for familiar space domain and also for this new time domain. The significance is that the Bohm Aharanov effect demonstrates that potentials have an effect on matter just as magnetic and electric fields do.

16.1 Introduction

In this paper, the electrodynamic potentials are evaluated with regards to the familiar 3D vector space and scalar time as well as 3D vector time and scalar space. The Bohm Aharanov effect demonstrates that vector potentials have real effects on electrons, shifting their phase. This means that potentials have effect on matter and should be included in the analysis of electrodynamics in three-dimensional vector time.

The concept of location in 3D vector space is very familiar to us. So is the concept of moving in 3D vector space. But to move in 3D vector space requires time. Without at least one dimension of time, it is not possible to move in space, since movement is change in space divided by change in time. A change in space divided by a change in space is not movement.

The concept of 3D vector time can be imagined, and we can even imagine moving from one location in time to another, just like in three-dimensional vector space. But when 3D vector time is put into mathematics, the idea of moving in time becomes an issue. A change in time divided by a change in time is not movement. Just like 3D vector space, to move in time requires a change in time and a change in space, specifically a change in time divided by a change in space, or the reciprocal of velocity in 3D vector space.

The speed of light is a ratio of space to time; increasing time has the same effect as decreasing space and vice versa. This reciprocal relationship between space and time will be used to couple space and time in a new way, where time is a three-dimensional vector and space is a scalar.

What does 3D vector time mean? We intuitively understand one dimension of time, but what about two dimensions of time [t^2], or even three dimensions [t^3]? Velocity and reciprocal velocity will be used to gain understandings of these new dimensions in time. Dewey Larson, in his Reciprocal System Theory, has raised some very interesting ideas, where the concept of three-dimensional time and the coupling of 3D vector time and 3D vector space via velocity[1]. In the Reciprocal System Theory, the lack of any mathematical derivation in the Reciprocal System can make it hard to follow how his concepts are derived. This paper proposes a mathematical approach to the idea of three-dimensional vector time with velocity as a coupler between space and time and looks at what impact this has on electrodynamics.

The form used in this paper are quaternions, the same form that Minkowski used to formulate spacetime with four dimensions, a three-dimensional vector for space and a scalar form of time. Quaternions were developed by W.R. Hamilton and were used by James Maxwell when he formulated his original Electrodynamic equations. In years after his work, Oliver Heaviside found that reducing a four-dimensional quaternion to a three-dimensional vector simplified the equations. The rationale for eliminating the scalar component is that the scalar portion had no physical analog, so could be ignored in the interest of simplicity. In this paper, as well as others to follow, the scalar portion has significance. The quaternion form that will be used in this paper is the complex quaternion, or bi-quaternion [2]. In this notation, all the scalars and vectors are complex. The bi-quaternion form of the quaternion is not explicitly required for the derivations in this paper, but I want to start with this format to allow maximum flexibility for future developments.

The Minkowski spacetime quaternion is significant because it couples space and time together into a 4D spacetime vector. It changed the way we looked at the universe. But is there still more to be gained from a reciprocal of this known spacetime quaternion, a new quaternion where the vector is a three-dimensional vector time and the scalar variable is space? These additional four dimensions of spacetime and how it applies to electrodynamics is the reason for this paper.

Note: In this paper, inverse and reciprocal mean the same thing.

16.2 Maxwell's equation in 3D Vector Time with SI units

For Maxwell's equations in SI units, the units of the electric field are now V/m and magnetic flux is V*s/ m^2. Permittivity (ε_s) is in units of A*s/V*m and permeability(μ_s) is in units of V*s/A*m, where and ε_s times μ_s equals $1/c^2$.

$\nabla_s \cdot E_s = - \rho_{es}/\varepsilon_s$ units of V/m^2 (1)

$\nabla_s \cdot B_s = 0$ units of Wb/m^3 or V*s/ m^3 (2)

$\nabla_s \times E_s + \partial B/\partial t = 0$ units of V/m^2 (3)

$\nabla_s \times B_s = \mu_s J_{es} + 1/c^2 \, \partial E/\partial t$ units of Wb/m^3 (4)

Using the new degrees of freedom, the proposed form of Maxwell's equations for 4D Time and 4D Space are the following: Equations relating to the electric field charge density and equation relating to magnetic field [B] have flux density.

For SI units, the following relations between vector space and vector time are important. This relationship is the only way the units work out and what the units suggest is very important.

$E_t = c^2 E_s$ units of Volt*meter/second^2 (5)

In vector time [4D Time], the electric field in time is related to the electric field in space by the relationship $E_t = c^2 E_s$ which converts V/m to V*m/s^2. This changes the electric field from one based on charge density to an electric flux.

$B_t = c^2 B_s$ units of Volts/second (6)

In 3D vector time [4D Time], the magnetic field in time has to be related to the magnetic field in space by the relationship $B_t = c^2 B_s$ which converts V*s/m^2 to V/s. This changes magnetic flux to a magnetic field based on charge density.

For permittivity and permeability,

$$\varepsilon_t = c^2 * \varepsilon_s = 1/\mu_s \qquad \text{units of A*m/V*s} \qquad (7)$$

$$\mu_t = c^2 * \mu_t = 1/\varepsilon_s \qquad \text{units of V*m/A*s} \qquad (8)$$

It is interesting to note that permittivity in 3D time becomes the inverse of permeability in 3D space. This is consistent with the electric field being a flux in 3D time. The same is true for permeability in time, which is the inverse of permittivity in 3D space.

$$\varepsilon_t * \mu_t = c^2 \qquad (9)$$

It is important to remember that since these next four equations describe electrodynamics in 3D Time, the basis units is per second instead of per meter as in 4D Space, since volume is measured in seconds instead of meters. Any movement in the scalar direction is time in 3D Space and space in 3D Time. The curl is taken with respect to time, so the space derivate is taken for the variable in the scalar dimension. Since magnetic monopoles are in 3D vector time, the units for the magnetic field are V/s^2, that is charge density, and the units for the electric field are in terms of flux so the units are V*m/s^2. In 3D Time, Maxwell's equations become,

$$\nabla_t \cdot E_t = 0 \qquad \text{units of V*m/s}^3 \qquad (10)$$

$$\nabla_t \cdot B_t = - \mu_t \, c \, \rho_{mt} \qquad \text{units are V/s}^2 \qquad (11)$$

$$\nabla_t \times E_t - c^2 \partial B_t / \partial s = c^2 J_{mt} / \varepsilon_t \quad \text{units are V*m/s}^3 \quad (12)$$

$$\nabla_t \times B_t + c\, \partial E_t / \partial s = 0 \quad \text{units of V/s}^2 \quad (13)$$

When magnetic charge density, or magnetic monopoles, along with an electrical flux, are introduced in 3D vector time, a symmetry is created with the electrical charge density and magnetic flux in 3D Space. In 3D vector space, ε_s is the electrical charge term, since electric fields have a source in charges and u_s is the magnetic flux. In 3D vector time, Electric fields are not derived from monopoles charges, and so E fields are treated as flux so ε_t is the electrical flux term and u_t the magnetic charge term.

Electric monopoles are measurable in 4D Space. But magnetic monopoles are not, just like three-dimensional vector time is not yet measurable. From these derivations, once one is detected, the other will also be detectable.

These eight electrodynamic equations give the full symmetry between space and time, impacting how electric and magnetic fields relate to these extra dimensions of spacetime. These additional degrees of freedom, as radical as they might currently appear, allow for magnetic monopoles from a four-dimensional field of 3D vector time and scalar space, where motion is always faster than light.

Movement in time manifests as energy, and acceleration in time manifests as a force. The familiar scalar time field is 3D vector time compressed into 3D scalar time. This scalar field must have extremely high uniformity since scalar time is used as a one-dimensional scalar variable without affecting the physics. It appears we have much more to learn about time. 3D vector time is easier to imagine, but movement in vector time requires scalar space, resulting in a reciprocal velocity from the perspective of 3D vector space. Just like the scalar time field manifests as a uniform scalar field, that is, as a uniform density in 3D space, so it could be imagined that scalar space field is a uniform density in three-dimensional vector time.

16.3 Vector and Scalar Potentials in 3D Vector Space and Scalar Time.

To derive the electrodynamic equations from potentials is done by starting with the equations[1]

$$\nabla_{\mathbf{s}} \cdot E_s = - \rho_{es}/\varepsilon_s \qquad \text{units of V/m}^2 \qquad (14)$$

$$\nabla_{\mathbf{s}} \cdot B_s = 0 \qquad \text{units of Wb/m}^3 \text{ or V*s/ m}^3 \qquad (15)$$

$$\nabla_{\mathbf{s}} \times E_s + \partial B_s /\partial t = 0 \qquad \text{units of V/m}^2 \qquad (16)$$

$$\nabla_{\mathbf{s}} \times B_s = \mu_s J_{es} + 1/c^2 \, \partial E/\partial t \qquad \text{units of Wb/m}^3 \qquad (17)$$

Since $\nabla_{\mathbf{s}} \cdot B_s = 0$, it must be the curl of some other vector. In electrodynamics, the vector chosen is A, which is the magnetic vector potential. It is represented as

$$B_s = \nabla_{\mathbf{s}} \mathbf{x} \; A_s \text{ [B field in terms of a potential]} \qquad (18)$$

From electrostatics that know that $\nabla_{\mathbf{s}} \mathbf{x} \; E_s$ is always 0, so the E field can be represented as the gradient of some scalar function. The E field is represented as

$$E_s = - \nabla_{\mathbf{s}}\Phi \qquad (19)$$

The first equation used to derive potentials is the curl of E with the partial derivative of B with respect to time replace by equation 19. This section follows the derivation done by Richard Feynman [2].

$$\nabla_{\mathbf{s}} \times E_s + \partial/\partial t \, (\nabla_{\mathbf{s}} \mathbf{x} \, A_s) = 0 \qquad (20)$$

$$\nabla_{\mathbf{s}} \times (E_s + \partial/\partial t \, A_s) = 0 \qquad (21)$$

The curl of a vector function that equals zero can be derived from the gradient of a scalar function,

In this case, the vector is set to the scalar function Φ. The same one in equation 19 is used,

$$E_s + \partial/\partial t\, A_s = -\nabla_s \Phi \qquad (22)$$

Now the electric field can be written in terms of the scalar and vector potentials.

$$E_s = -\nabla_s \Phi - \partial/\partial t\, A_s \text{ [E field in terms of potentials]} \qquad (23)$$

In order to define both the electric field and magnetic field in terms of potentials, we need four equations for the potentials.

It is possible to define a new potential function that does not change the physics of the E and B fields.

$$A' = A + \nabla_s \chi \qquad (24)$$

and

$$\Phi' = \Phi - \partial/\partial t\, \chi \qquad (25)$$

Now the objective is to determine A and Φ from the equations with ρ and J. Using the divergence of the electric field

$$\nabla_s \cdot E_s = -\rho_{es}/\varepsilon_s \qquad (26)$$

and using the E field from equation 23

$$\nabla_s \cdot (-\nabla_s \Phi - \partial/\partial t\, A_s) = -\rho_{es}/\varepsilon_s \qquad (27)$$

$$\nabla^2_s\Phi \;\; - \partial/\partial t\, (\nabla_s \cdot A_s) = - \rho_{es}/\varepsilon_s \tag{28}$$

So we have charge density in terms of scalar and vector potential, but it is not complete because there is still the divergence of A term.

Using the last Maxwell equation (17)

$$\nabla_s \times B_s - 1/c^2\, \partial E/\partial t = \mu_s\, J_{es} \tag{29}$$

and substitute in the magnetic vector potential and E from equation 23

$$\nabla_s \times (\nabla_s \times A) - 1/c^2\, \partial/\partial t\, (- \nabla_s\Phi \;\; - \partial/\partial t\, A_s) \;\; = \mu_s\, J_{es} \tag{30}$$

Using the vector identity

$$\nabla_s \times (\nabla_s \times A) \; = \nabla_s\, (\nabla_s \cdot A) - \nabla^2_s A \tag{31}$$

and applying this vector identity to equation 30

$$-\nabla^2_s A + \nabla_s\, (\nabla_s \cdot A) + 1/c^2\, \partial/\partial t\, \nabla_s\Phi + \; 1/c^2\, \partial^2/\partial t^2\, A = \mu_s\, J_{es} \tag{32}$$

Again, there is the option to choose an arbitrary divergence of A,which still satisfies the physics of E and B.

$$\nabla_s \cdot A = - 1/c^2\, \partial/\partial t\, \Phi \tag{33}$$

Inserting equation 33 into 32

$$-\nabla^2_s A - \cancel{\nabla_s (1/c^2\, \partial/\partial t\, \Phi)} + \cancel{1/c^2\, \partial/\partial t\, \nabla_s \Phi} + 1/c^2\, \partial^2/\partial t^2 A = \mu_s J_{es} \qquad (34)$$

The second and third term cancel, so the equation simplifies to

$$\nabla^2_s A - 1/c^2 \partial^2/\partial t^2 A = -\mu_s J_{es} \qquad (35)$$

Now we go back and insert the definition of the divergence of A into equation 28:

$$\nabla^2_s \Phi - 1/c^2\, \partial^2/\partial t^2\, \Phi = -\rho_{es}/\varepsilon_s \qquad (36)$$

So now all Maxwell's equations defined in terms of scalar and vector potentials

$$E_s = -\nabla_s \Phi - \partial/\partial t\, A_s \qquad (37)$$

$$B_s = \nabla_s \times A_s \qquad (38)$$

$$\nabla^2_s \Phi - 1/c^2\, \partial^2/\partial t^2\, \Phi = -\rho_{es}/\varepsilon_s \qquad (39)$$

$$\nabla^2_s A - 1/c^2\, \partial^2/\partial t^2 A = -\mu_s J_{es} \qquad (40)$$

In free space, $\rho_{es} = 0$, equation 39 becomes the scalar wave equation

$$\partial^2/\partial x^2\, \Phi = 1/c^2\, \partial^2/\partial t^2\, \Phi \qquad (41)$$

16.4 Vector and Scalar Potentials in 3D Vector Time and Scalar Space

In vector time and scalar space, the electric field is a flux, so is derived from an Electric Vector potential F_t and the magnetic field is derived from a scalar potential ψ_t. In vector space, the units of scalar potential is volts and the magnetic vector potential is V*s/m. In vector time, the scalar potentials is still volts, but the electric vector potential becomes V*m/s.

$$E_t = - \nabla_t \times F_t \qquad \text{units of } V^*m/s^2 \qquad (42)$$

and

$$B_t = \nabla_t \psi_t \qquad \text{units are } V/s^2 \qquad (43)$$

In Electrodynamics, the equation for the electric field and the electric vector potential, F, is defined as $E_s = \nabla_s \times F$. For the magnetic field and the scalar magnetic potential ψ, the equation is $B_s = \nabla_s \psi$. But these do not work for 3D vector time because the del operator for 3D time is different than for 3D space and the units are in terms of time for area and volume and time/space for motion. In the space domain, the vector potential A_s has units of V*s/m, but in the time domain, F_t has units of V*m/s. The units for Φ and ψ_t are the same, both V.

The starting electrodynamic equations for 3D time are [1],

$$\nabla_t \cdot E_t = 0 \qquad \text{units of } V^*m/s^3 \quad (44)$$

$$\nabla_t \cdot B_t = - 1/c^2 \mu_t \rho_{mt} \qquad \text{units are } V/s^2 \quad (45)$$

$$\nabla_t \times E_t - c\, \partial B_t / \partial s = c\, J_{mt} / \varepsilon_t \quad \text{units are } V^*m/s^3 \quad (46)$$

$$\nabla_t \times B_t + \partial E_t / \partial s = 0 \qquad \text{units of V/s}^2 \qquad (47)$$

Beginning with equation 47 and inserting equation 42, the results are

$$\nabla_t \times B_t + \partial / \partial s \, (-\nabla_t \times F_t) = 0 \qquad (48)$$

$$\nabla_t \times (B_t - \partial / \partial s \, F_t) = 0 \qquad (49)$$

As before, a vector function that has a curl of zero can be defined as the gradient of a scalar function. Using the same logic the previous section,

$$B_t - \partial / \partial s \, F_t = -\nabla_t \psi_t \qquad (50)$$

Rearranging this equation,

$$B_t = -\nabla_t \psi_t + \partial / \partial s \, F_t \qquad (51)$$

Again, redefining the vector and scalar potentials so they do not change the physics of the E and B fields.

$$F_t' = F_t + \nabla_t \chi \qquad (52)$$

$$\psi_t' = \psi_t - \partial / \partial s \, \chi \qquad (53)$$

$$\nabla_t \cdot B_t = -1/c^2 \mu_t \rho_{mt} \qquad (54)$$

$$-\nabla_t \cdot (\nabla_t \psi_t - \partial / \partial s \, F_t) = -1/c^2 \mu_t \rho_{mt} \qquad (55)$$

$$\nabla^2_t \psi_t - \partial / \partial s \, (\nabla_t \cdot F_t) = 1/c^2 \mu_t \rho_{mt} \qquad (56)$$

To complete the derivation, we need to solve the last electrodynamic equation in time.

$$\nabla_t \times E_t - c^2 \partial B_t / \partial s = c J_{mt} / \varepsilon_t \qquad (57)$$

And substitute in equations for E_t and B_t

$$\nabla_t \times (-\nabla_t \times F_t) + c^2 \partial/\partial s \nabla_t \psi_t + c^2 \partial^2/\partial s^2 F_t = c J_{mt} / \varepsilon_t \qquad (58)$$

Using the vector identity

$$\nabla_t \times (-\nabla_t \times F) = -\nabla_t (\nabla_t \cdot F) + \nabla^2_t F \qquad (59)$$

Applying this vector identity in equation 58

$$\nabla^2_t F_t - \nabla_t (\nabla_t \cdot F) - c^2 \partial/\partial s \nabla_t \psi_t - c^2 \partial^2/\partial s^2 F_t = c J_{mt} / \varepsilon_t \qquad (60)$$

Again, there is the option to choose an arbitrary divergence of A, which still satisfies the physics of E and B.

$$\nabla_t \cdot F_t = - c \partial/\partial s \psi_t \qquad (61)$$

Now insert equation 61 into 60

$$\nabla^2_t F_t + \cancel{\nabla_t (c \partial/\partial s \psi_t)} - \cancel{c \partial/\partial s \nabla_t \psi_t} + c^2 \partial^2/\partial s^2 F_t = c J_{mt} / \varepsilon_t \qquad (62)$$

The second and third term cancel, so the equation simplifies to

$$\nabla^2_t F_t + c^2 \partial^2/\partial s^2 F_t = c J_{mt} / \varepsilon_t \qquad (63)$$

Now we go back and insert the definition of the divergence of F_t into equation 56.

$$\nabla^2_t \cdot \psi_t - c\, \partial/\partial s\, (c\, \partial/\partial s\, \psi_t) = 1/c^2\, \mu_t\, \rho_{mt} \tag{64}$$

which simplifies to

$$\nabla^2_t \cdot \psi_t - c^2\, \partial^2/\partial s^2\, \psi_t = 1/c^2\, \mu_t\, \rho_{mt} \tag{65}$$

So now we have all the equation defined in terms of scalar and vector potentials.

$$E_t = - \nabla_t \times F_t \tag{66}$$

$$B_t = -\nabla_t \psi_t + c\, \partial/\partial s\, F_t \tag{67}$$

$$\nabla^2_t \cdot \psi_t - 1/c\, \partial^2/\partial s^2\, \psi_t = 1/c^2\, \mu_t\, \rho_{mt} \tag{68}$$

$$\nabla^2_t F_t + \partial^2/\partial s^2\, F_t = c\, \mu_s\, J_{es} \tag{69}$$

16.5 Wave equation

In free space, the wave equation for the space domain, vector space and scalar time, is shown to be

$$\partial^2/\partial x^2\, \Phi = 1/c^2\, \partial^2/\partial t^2\, \Phi \tag{70}$$

In free space, $\rho_{mt} = 0$, so equation 68 becomes the scalar wave equation

$$\partial^2/\partial t^2\, \psi_t = c^2\, \partial^2/\partial x^2\, \psi_t \tag{71}$$

Rearranging this equation

$$\partial^2/\partial x^2\, \psi_t = 1/c^2\, \partial^2/\partial t^2\, \psi_t \tag{72}$$

This is the exact same equation as vector space and scalar time. This seems to suggest that potentials are common to both.

This is also true for the vector potential. For the space domain, it is

$$\partial^2/\partial x^2 A_s = 1/c^2 \, \partial^2/\partial t^2 A_s \tag{73}$$

For the time domain, it can be rearranged to be

$$\partial^2/\partial x^2 F_t = 1/c^2 \, \partial^2/\partial t^2 F_t \tag{74}$$

So the format for the vector potential equations are exactly the same, but the vector potential is not. The units are identical, so the scalar potentials are the same. This is not true for the vector potentials, where it was noted previously that the units for the vector potentials are different.

As is typical, these two vector potentials are coupled by c, where $F_t = 1/c^2 A_s$. For the paper 'Units for Electrodynamic Equations in Three-Dimensional Time,' it can be seen that only the speed of light constant is needed to convert properties between the space and the time domain [3].

16.6 Bohm Aharonov Effect.

The Bohm Aharonov effect links the behavior of the electron not with electric and magnetic fields but with potentials. This effect was proposed in 1959 to show that the magnetic vector potentials is not just a theoretical construct, but has physical effects [4]. The first experiment was run in 1960 and showed a phase shift in electron interference fringes. Due to the effect that even a tiny residual magnetic field can have on an electron, the experiment was run again in 1986, this time with the latest material shielding technologies. Again, the same effect is seen. In the latter case, a superconductor was used and the magnetic flux is quantized, which quantized the phase shift seen in the electron interference pattern

$$\Delta\Phi = (e_0\ \Phi_m / \hbar) = \pi n \tag{75}$$

In the experiment, two beams of electrons go around opposite sides of a solenoid to a screen where interference fringes can be imaged. The two different paths are identical. Outside the solenoid, the magnetic field is zero. When the magnetic field is turned on, the interference fringes of the electron shift, since the interaction of potentials results in a phase shift.

The same experiment was done with scalar potentials, where different potentials were applied to cylindrical metal tubes in each path. Again, the same result.

The Aharonov and Bohm effect is not limited to the magnetic vector potential. Aharonov and Bohm also describe a situation where particles are affected by regions where the electric field is zero, an electric Aharonov Bohm effect [5]. ***So both the scalar potential and vector potentials affect particles without fields.***

In their paper, Aharonov and Bohm state that potentials are typically regarded as mathematical constructs and are considered to not have physical significance because equation of motions only involve fields. They, and other authors, after summarizing the results of the Aharonov and Bohm effect, believe potentials to be more fundamental than fields [4,5,6].

16.7 Summary

The electrodynamic equations for three-dimensional vector time have been derived based on the equations in 'Concepts of Three-dimensional Time in Electrodynamics [1]. The units for each step have been checked to make sure that they are consistent with Three-Dimensions of vector time and motion in vector time [3].

Evaluating electrodynamics in terms of scalar and vector potentials show that the scalar wave equation is exactly the same for the space domain (vector space and scalar time) as it is for the time domain (vector time and scalar space). This suggests that the scalar potential is common to both the space and the time domain. The units for this scalar potential remain volts in both domains.

The vector potential has the same equation format as well, but the vector potential between the space and time domains is different by a factor of $1/c^2$. In the space domain, this generates the magnetic field, and in the time domain it generates the electric field.

The symmetry of scalar and vector potentials and the scalar and vector terms for the space and time domain might be coincidence, but on the other hand, they might lead to some deeper understanding of the properties of time in the near future.

References

1. Concepts of Three-Dimensional Time in Electrodynamics, Robert Kersten, www.multidimensionaltime.com
2. The Feynman Lectures, Volume II, section 18-6.
3. Units for Electrodynamic Equations in Three-Dimensional Time, Robert Kersten, www.multidimensionaltime.com
4. Significance of Electromagnetic Potentials in the Quantum Theory, Y. Aharonov and D. Bohm, The Physical Review, Vol. 115, No. 3, Aug 1, 1959.
5. Aharonov-Bohm effect, Ambroz Kregar, Department of Physics, Univerza v Ljubljani, marec 2011. http://mafija.fmf.uni-lj.si/seminar/files/2010_2011/seminar_aharonov.pdf
6. Aharonov-Bohm effect and Magnetic Monopoles, bolvan.ph.utexas.edu/~vadim/Classes/2011f/abm.pdf

Version

1. Original
2. Updated with reference to Electric Vector Potential, **F**, and magnetic scalar potential ψ. Changed the section in 3D time so that A_t is now F_f and Φ_t is changed to ψ_t so that it correlates better to the electric vector and magnetic scalar potential in electrodynamics. Add in reference four. Note: in other sources, the electric vector potential is denoted by vector C.
3. Included Aharonov-Bohm effect.

17 A 3 – Three-dimensional Time and Zero Point Energy

Robert Kersten
www.multidimensionaltime.com

Dec 29, 2014

Abstract.

In a previous paper, Concepts of Three-dimensional Time in Electrodynamics, the concept of three-dimensional vector time and scalar space is introduced along with its impact to electrodynamics. In that paper the properties of 3D vector time are compressed into a scalar time field and the properties of 3D vector space are compressed into a scalar space field. Both scalar fields are also three-dimensional, but have very high uniformity, so time can be considered one-dimensional without impacting physics. In this paper, we will explore the properties of multidimensional time, primarily in its more familiar scalar form. In doing so, the similarity between the physics of Stochastic Electrodynamics and this scalar field of time becomes apparent. Three-dimensional vector and scalar time, as well as three-dimensional vector and scalar space are coupled via motion, specifically velocity. Stochastic Electrodynamics [SED] is based on a motion, a jitter motion or zitterbewegung. This motion is referred to as Zero-Point Energy, and it couples the Zero-Point Field to matter. It is from this basis that I will expand the understanding of 3D and scalar time physics and show the commonality it has with Zero-Point Energy.

17.1 Introduction

In 1901, Max Plank published his first paper on the radiation of a black body. In his paper, Plank introduced a new constant, h, which came to be known as Plank's constant, to accurately calculate the radiation emitted from a black body. What was disturbing at the time was that the radiation had to be quantized. This quanta of energy is small, but the field of physics up to that time believed energy to be a continuous phenomenon. At the macro level, this has little impact, but at the microscopic level, specifically the level of atoms and particles, it has a radical impact. Plank was not comfortable with the format of his equation. He worked on it for another ten years and then published a new paper in 1911. In this second paper, he had two terms instead of one in his equations, as shown below. In the first term in this equation, energy is dependent on temperature, and in the second term it is not dependent on temperature.

$$E = (h\nu/e^{[h\nu/kT]}-1) + h\nu/2 \qquad (1)$$

In this equation, h is Plank's constant, ν is frequency, k is Boltzman's constant, and T is absolute temperature. The result of this equation is that matter always has a residual amount of energy, even at zero degrees. Because of this, this second term became known as the Zero-Point Energy [ZPE].

The concern with this term was that although the energy generated by each individual frequency is extremely small, the universe is expected to support an infinite number of frequencies, so the amount of energy density from all these frequencies is infinite. To solve this problem, the frequencies are limited to the point where the fabric of space breaks down, Plank's length. This frequency, which is on the order of 10^{44} Hz, is inserted into the equation 1, and the amount of energy density is recalculated. The result is still extremely huge, an energy density on the order of 10^{93} g/cm^3. Atomic energy, the most dense form of energy we knew before this, only has an energy density of 10^{14} g/cm^3. This energy density is so large, it has the capacity to vaporize all the known matter in the universe. The reason that the ZPE does not destroy matter is that it is extremely uniform, so there are no forces to create this destruction [1]. It is like a box that is closed and sealed at sea level. So at sea level, there is uniform pressure inside or outside the box. But if the box is taken deep under the ocean, it will be crushed by the outside force generated by the in-balance of pressures outside versus inside the box. Likewise, if the box is brought into the upper part of the atmosphere, it will explode because the force generated by the larger pressure inside the box. So the uniformity is critical to the balance of the universe as we know it.

17.2 Stochastic Electrodynamics [SED].

Out of this second equation, a second interpretation of quantum behavior is developing, a field called Stochastic Electrodynamics [SED]. SED postulates that the Zero Point Energy [ZPE] is an energy that manifests from a gradient, or change in the balance of the Zero-Point Field. As noted above, it is the difference in potentials, or pressure in the analogy, that creates the energy and forces. This ZPE is a jitter motion that is imparted to matter at the most fundamental level. The Zero Point Energy consists of virtual particle pairs, in the form of electron-positron pairs, that oscillate between an undetectable presence in this Zero-Point Field and a very brief detectable presence as photons in the universe of matter. During the brief time as photons, these photons can absorb and emit electromagnetic radiation, hence interact with matter. At a given ZPE energy density, the amount of virtual particle density is fixed. This virtual particle density fixes the amount of energy imparted to real particles during the brief interaction.

H. E. Putoff asked the question why all the electrons have not run out of energy and crashed into the nucleus of the atom, since, as they orbit the nucleus, they are constantly radiating energy. He calculated that the electron has the orbit radius it has because the electron absorbs just the right amount of energy from the ZPF to maintain this exact stable orbit. If the energy density of the ZPE decreases, the electron's path will get smaller, and the electromagnetic attraction of the nucleus might slowly pull the electrons in, unless it absorbs enough energy to counterbalance this electromagnetic attraction. Likewise, if the energy density of the ZPE increases, the orbit of the electrons will increase, and they could escape the electromagnetic attraction, or coulomb force, of the nucleus.

Since the atomic frequencies detected by our instruments are dependent on the radius of the electrons' orbit, and the ZPE affects the orbits of the electrons in all matter, this means that the energy density of the ZPE impacts the precision of the atomic clocks we use.

17.3 Three-dimensional Time

In the paper 'Concepts of Three-dimensional Time in Electrodynamics,' the idea of dual symmetry, where electric fields are derived from electrical monopoles and magnetic fields are derived from magnetic monopoles was expanded to include three dimensions of a vector time field and a scalar space field [1]. The idea of three-dimensional time and scalar space means that length, area, and volume are in terms of seconds, seconds^2 and seconds^3 and movement in time is in terms of meters. In the paper, it was explained that movement in 3D Space requires time, that is, dx/dt, and movement in 3D Time requires space, represented as dt/dx, or inverse velocity. So the units of electrodynamics in 3D Time has to reflect this fact. When Maxwell's equations are adapted to 3D Time, in order to get the units correct, some interesting phenomena become clear, as the equations in that paper showed.

In 3D Space, electric fields are sourced from electric monopoles, and the magnetic fields are a flux without monopole sources. It lands up that the electric field is a flux in 3D Time, and the magnetic field is derived from monopoles that exist in 3D Time, the opposite of 3D Space. So the symmetry of monopoles and flux is captured with these additional dimensions of time and space. In addition, velocity in 3D Time is always faster than the speed of light from the perspective of 3D Space. Any time scalar time or scalar space is referred to, it is implicit that it is a three-dimensional scalar field that is treated as a one-scalar variable due to extreme uniformity. For clarity, the definition of a scalar field is a field that only has a single value for each location in that field. A vector field, by comparison, has a value or magnitude and a direction for each location in the field.

The following was captured in the summary of the above-mentioned paper:

"*These eight electrodynamic equations give the full symmetry between space and time, impacting how electric and magnetic fields relate to these extra dimensions of spacetime. These additional degrees of freedom, as radical as they might currently appear, allow for magnetic monopoles from a four-dimensional field of 3D vector time and scalar space, where motion is always faster than light.*

Movement in time manifests as energy, and acceleration in time manifests as a force. The familiar scalar time field is 3D vector time compressed into 3D scalar time. This scalar field must have extremely high uniformity, since scalar time is used as a one-dimensional scalar variable without affecting the physics. It appears we have much more to learn about time. 3D vector time is easier to imagine, but movement in vector time requires scalar space, resulting in a reciprocal velocity from the perspective of 3D vector space. Just like the scalar time field manifests as a uniform scalar field, that is, as a uniform density in 3D Space, so it could be imagined that scalar space field is a uniform density in three-dimensional vector time."

To understand time better, especially three-dimensional vector time, it is helpful to compare it to three-dimensional vector space. In three-dimensional vector space, we generally use the earth's surface as a reference for location. We can identify a location on earth, in terms of, let's say for this analogy, x, y, and z coordinates and label it the origin. We can move away from that location in any direction and then return to that exact location at any future time. As long as it is referenced to the earth, it is an absolute location. If we reference it with respect to our sun, it is not the same location anymore since in the time we moved away and back to this origin on earth, the earth has moved in its orbit around the sun. But if you wait long enough, you can get to the same location in the orbit again. If you reference this location relative to the galaxy, it gets more complicated. So with space, you can normalize your position coordinate system so you can return to an absolute location.

But this is not how it is for time. For this analogy, let's assume we have three dimensions of vector space and scalar time, our typical experience of spacetime as we currently define it. If we mark time just as we move out of the origin, walk in any direction and then come back, we are at the same location in space, but we can never go back to the same time. At every moment, time is changing constantly. We can never go back in time, physically. So this three-dimensional scalar time field is always increasing in magnitude, no matter what direction you choose. The first time measurement becomes t=x seconds, and then we measure from this moment. This method is the same, no matter what clock is used and when that clock started. The clock could be celestial, solar, atomic, it makes no difference. All we can do in time is measure a change, a delta. When motion is involved, we measure a change in space and divide it by this change in time [dx/dt]. All the directional properties of motion are captured in space, since time has no directional properties when physics is done in 3D vector space and scalar time.

To complete this analogy, let's move into a new neighborhood of spacetime, where time is three dimensions and has the property of direction in all three dimensions. We can measure any location in terms of seconds. The location can be defined in terms of t_x, t_y and t_z. It is possible, just like space, to move away from that location and return to it. It is well-defined. But now space is scalar, and like scalar time, it is always increasing in magnitude. To a time traveler in this neighborhood, it has the same issues as scalar time has to a space traveler. In this neighborhood, you can leave and come back to the same location in time, but space is always changing. To measure motion, you take a change in time and divide it by a change in space, that is, dt/dx or the inverse of space velocity. So space becomes the variable that can only be measured in terms of a change, or delta. In the paper referenced in the beginning of this section, it was shown that the properties related to three-dimensional vector time are energy and force. In regards to scalar time that we are familiar with, these properties can only manifest in a scalar form. The only form of energy we know of is a scalar form which is independent of direction and so can be a property of this three-dimensional scalar time field. But force as it is currently defined is not a scalar. Energy is defined as a change in time divided by a change in space.

Our human vision is a good analogy of how 3D vector time is undetectable and scalar time is detectable. We 'see' because our eyes are in constant motion, measuring and recording differences. When an experiment was done to compensate for the motion of a subject's eyes, they could not 'see' anymore. This same effect makes the fundamental field of three-dimensional time undetectable to us, since we need motion,or change, to detect time.

17.4 Time fields and Zero-Point Energy

Zero-Point Energy and the Zero-Point Field are related as follows: The Zero-Point Field is potential energy, but very uniform potential energy density. Zero-Point Energy is the motion that results from differences, or asymmetries, in the Zero-Point Field potentials. This Zero-Point Energy is a scalar field, since the motion is in all directions, so no particular direction can be associated with this energy field.

There are four properties of the ZPF that are typically associated with this field. First is the property of undetectability, the second is extreme uniformity, third is it is everywhere in space, the fourth is energy density. ZPF is undetectable due to the fact that it is Lorentz invariant, so constant motion through the field doesn't make it detectable, but accelerated motion does make it detectable, and it is extremely uniform and is everywhere.[2].

ZPE has five properties that are important. First is the property of detectability, the second is uniformity, the third is kinetic motion, the fourth is density, and the last one is that it is a scalar field. ZPF becomes detectable when its potential energy is converted into kinetic energy as a motion, called Zero-Point Energy. We measure this kinetic energy of the Zero-Point Energy as a frequency, which is period of time, and a period is just a change in time, so Zero-Point Energy shares this same measurement with scalar time.

Now the properties of the scalar time field are as follows: It is everywhere, is extremely uniform, is a scalar field, and has the property of energy. The property of energy comes from the fact that energy is mass, and mass is a time phenomenon, as discussed in the next section [1].

The field of time does share extreme uniformity that makes it undetectable. Time's uniformity comes from the fact that we can treat it as a single value in a three-dimensional scalar field. *It is important to point out that it is the field of time that is undetectable, not the difference in the field of time, which is what is measured as time in three-dimensional space. Just as ZPF is undetectable, but the difference in ZPF, which manifests as ZPE, is detectable.* Our measurement of time is coupled to velocity, so there is no way to measure time without a reference to space. With time, we are able to measure the change in location of time, but we cannot measure ***a location*** in time independent of motion. It is these locations in time that make up the 3D vector time field, which is undetectable. We can measure the kinetic results of ZPE on matter, but we cannot measure the potentials of ZPF.

So location in time is undetectable to us, but we have managed to use the change in time locations as our measure of time very effectively. But as shown in my first paper, multidimensional time has very important information that we are not using due to our limited view of time, just as the concepts in Zero-Point Energy are opening our minds to new possibilities in physics [1]. Potentials do not require motion, but they do require location. Gravitational potentials are purely the potential difference of different locations. For

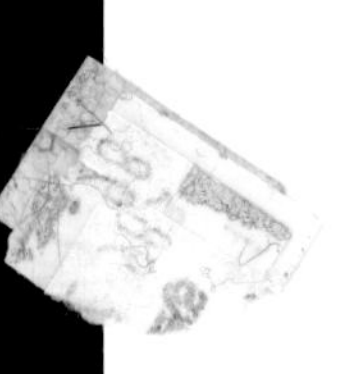

electromagnetic potentials, if no gradient is possible, no electric or magnetic fields are possible. Just like gravity is a potential generated by location, so time has potential generated by locations in the field of time. A simple analogy would be the probability, or potential, of a particular outcome in the future has different values for different vectors in this time field. These potentials have gradients, and therefore field densities, associated with them. Where these gradients manifest in the 3D time field, scalar energies should manifest in the Scalar Time Field just like ZPE. So motion is embedded in the very fabric of matter and is the generator of spacetime. Velocity is the ratio of the amount of space to the amount of time. In our part of the measurable universe, this ratio is 299 million kilometers for every second of time.

With a scalar time field, asymmetries in the undetectable time potentials of the 3D time field generate the energy density associated with that field. In addition, I would expect that there would be a similar cutoff variable in this process, probably a frequency in time that limits the amount of energy density that is manifested in the Scalar Time Field. In the physics of ZPE, if the energy density of the ZPE increases by more than a quantum of energy, every single electron in the universe will shift in its energy at the exact same time [3]. Like the energy density of the ZPE, the density of the Scalar Time Field would have the same function, a universal time keeper for all of matter at the most fundamental level, with frequency being a critical measurable variable of the 3D time field.

The speed of light is a scalar space field ratioed to scalar time field, generally uniform in all directions. One very interesting, and frustrating, fact is that the speed of light, being a scalar value, cannot be used to couple the location in 3D Time field to a location in 3D Space field. Hopefully a better understanding of the vector time and space fields in the future will allow this ability.

The difference here is that ZPF is a scalar field, and three-dimensional time is a vector field, but directions which are undetectable to us. For now, the best correlation of the ZPF is to the scalar potentials, which are common to both 3D Space and scalar time and to 3D Time and scalar space. [7]

17.5 Mass

In an article called Zero Point Energy, by the Calphysics Institute, their scientists remind us that the Higgs field does not explain the origin of inertial mass of ordinary matter. The article states: "*The Higgs field applies only to the electro-weak sector of the Standard Model. The mass of ordinary matter is overwhelmingly due to the protons and neutrons in the nuclei of atoms.*" In the same article, it states, "*The origin of inertial mass of ordinary matter is thus a wide-open question.*"[4]

In the Reciprocal System Theory [5], mass is defined in terms of inverse velocity. The theory postulates, and as shown in the paper 'Concepts of Three-dimensional Time in Electrodynamics,' motion in space is not possible without time. But the theory takes this a step further. Motion in 3D Space is not possible with matter that is part of 3D Space. There must be time involved, that is, a dx/dt. If matter is moving in 3D Space, it must have a footing in 3D Time. In the equation F=ma, mass is represented as $1/c^3$, that is $time^3/space^3$, a unit of 3D Time. In this same paper, inverse velocity cubed is used as a coupler between force and acceleration.

In papers on ZPF and ZPE, the authors have worked out formulations to show that mass is due to a reaction of the ZPF to accelerated motion in the field of ZPF, which is everywhere. This reaction to accelerated motion of charge is inertia, which is measured as inertial mass in vector space and scalar time. So mass is not an innate property of matter, as theory states. In the ZPF paper "Beyond $E=mc^2$", the authors state: "*In our formulation, the m in Newton's second law of motion, F=ma, becomes nothing more than a coupling constant between acceleration and an external electromagnetic force.*"[2]. In my paper, Energy was shown as a property of the 3D Time Field with units of time/space, or 1/c, as postulated by the Reciprocal System Theory. If the equation $E = mc^2$ is solved for mass, then mass = E/ c^2, and E has been shown to be equal to 1/c, so mass is $1/c^3$. So mass is energy in three dimensions of time. Again, in the same paper "Beyond $E=mc^2$", the authors state the following, based on the derivation of mass from ZPF: "*In the view we will present, Einstein's formula is even more significant than physicists have realized. It is actually a statement about how much energy is required to give the appearance of a certain amount of mass, rather than about the conversion of one fundamental thing, energy, into another fundamental thing, mass.*" Later in the same paper, they are more explicit: "*Mass is energy.*" [2] The 3D Time derivation of mass says exactly the same thing, where energy is a property of time. So mass is not an innate space property of matter; it is a result of energy generated by motion through the field of time. [5] In addition, the Reciprocal System Theory states that motion in time is in opposition to motion in space. The theory of ZPE and time appear to correlate pretty well.

17.6 Scalar Time Field, Vector Time Field, and Dark Energy

Physicists have proposed that while the frequency limit for the Zero-Point Field can be as high as Plank's frequency, there is likely a cutoff frequency at which density of ZPE is limited to. One approach to estimating a cutoff frequency is to estimate the energy density and solve for the cutoff frequency. If the energy density of visible and dark energy is taken into account, and then this equation is solved to determine an upper limit of frequency, the result is in the low Tera Hertz [10^{12}] range [6].

Cosmologist found that in order to explain the expanding universe, there has to be matter that is creating a negative pressure, or expanding force, that accounts for this expansion. Since the energy cannot be detected, it is labeled dark energy.

Beck et al, have proposed that with a cutoff frequency in the low THz range, the energy density of ZPE matches that of the energy density of dark energy. They postulate that the ZPE could be the source of this dark energy as well, since it manifests as a negative pressure, causing the expansion of the universe. [6]

It does not seem unreasonable to use this same idea of a cutoff frequency, since frequency is a property of time, as a limit of how much of the 3D Time Field manifests in the Scalar Time Field. In this way, the density of time in the scalar time field is determined, and as the density of time changes, so does the timing [frequency] and density of matter.

As mentioned in the previous section, the Reciprocal System Theory states that motion in time is in opposition to motion in space. So if motion in time is toward all other locations, motion in space would be away from all other locations. So a contraction in vector time and scalar space manifests as an expansion in vector space and scalar time. An expansion of our universe is what astrophysicists are currently measuring, and they postulate this Dark Energy as the explanation for this expansion [5].

If the expansion of matter in vector space and scalar time is driven by a contraction of matter in vector time and scalar space, it could mean that gravity is associated with vector time and scalar space, and it is this gravity force in time that is driving the expansion of matter in space. The forces are opposite to what physics currently defines it, where unobservable gravity is in vector space and scalar time and this unobservable Dark Energy is somewhere in vector space and scalar time as well. The Reciprocal System Theory already has this expansion and contraction of matter built into the theory, so it is a natural fit for any force that is driving the expansion of our universe [Dark Energy] as well as the force that is driving the contraction of matter, that is, gravity. What this postulate of gravity in time being the driver of expansion in space does is it links together gravity, time, and magnetism. Time and magnetism are already correlated as shown in the paper 'Concepts of Three-Dimensional Time in Electrodynamics' [1].

The extreme uniformity of the ZPF, and of Time, means that there is a Balance. Without the balance, the gradients would manifest force that would rip matter as we know it apart. For every action, there is an equal and opposite reaction. The model of an eight-dimensional vector space and scalar time and vector time and scalar space allows this model opposite reaction.

17.7 Conclusion

This paper focused on the common properties between time and zero-point energy. In this paper, as well as in my previous papers, I derive and delineate some properties of Time postulated by the Reciprocal System Theory, but include new properties as well. There are a number of properties of 3D Scalar Time fields that are shared with Zero-Point Energy, such as undetectability, uniformity, motion, density, and frequency. In addition, the ZPE is the universal time keeper for all matter, a property associated with time, so it is plausible that these time fields are the same as ZPE. Just as a cutoff frequency limits how much Zero-Point Energy density manifests out of the Zero-Point Field, so too a cutoff frequency could be applied the amount of time scalar density that manifests out of the 3D time vector field. A more dense time field would mean the locations in the time field are closer together, and since we use the change in time locations as our measurement of time, it would change time as we measure it.

The postulate of the Reciprocal System Theory that motion in time is in opposition to motion in space could account for the energy, labeled dark energy, and this opposition is the expansion that is measured in our universe. This correlates time to gravity, and in an earlier paper, time to magnetism. Hopefully in the future, we can uncover what this relationship might be.

The origin of mass is not well-known, and Stochastic Electrodynamics has provided a very interesting insight that mass is the coupling between force and acceleration and mass ***is*** energy, exactly as it is in the 3D Time [5].

The field of Stochastic Electrodynamics provides a new way of looking at our universe, as does the dimensions of 3D Time and scalar space. I believe the overlap of these two theories, multidimensional time and Stochastic Electrodynamics, will give us new insights into spacetime that were not available before. It is possible that viewing ZPE as a time phenomenon could lead to faster insights into dimensions of spacetime beyond the four we currently work with.

References

1. Concept of Three-Dimensional Time in Electrodynamics, Robert Kersten, www.multidimensionaltime.com
2. Beyond E=mc^2, Bernhard Haisch, Alfonso Rueda, H.E. Putoff, The Sciences, Vol. 34, no. 6, p 26-31, Nov/Dec 1994.
3. Cosmology and the Zero-Point Energy, Barry Setterfield, Natural Philosophy Alliance Monograph Series, No.1, 2013, p16.
4. Zero-Point Energy, Calphysics Institute, http://www.calphysics.org/zpe.html
5. *Nothing But Motion*, Dewey Larson, North Pacific Publishers, 1979. ISBN 0-913138-07-x. Information can also be accessed at www.reciprocalsystem.com/dbl/ .
6. Measurability of vacuum fluctuations and dark energy, Christian Beck, Michael C. Mackey, Dec 11, 2006, http://arxiv.org/pdf/astro-ph/0605418.pdf
7. Electromagnetic Potentials in Three-Dimensional Time, Robert Kersten, www.multidimensionaltime.com

Version

1. Original
2. Jan 03,2014. Update correlation of ZPF to scalar potentials, section 4.
3. Nov 30, 2015. Update to include gravity as a time phenomenon, as well as the link of time to magnetism.

18 Index

Made in the USA
Lexington, KY
04 April 2016